MW00416749

ISBN 978-0-265-77778-7
PIBN 10888085

International Atlas

of

Clouds

and of

States of the Sky

THIS WORK
FOR THE USE OF OBSERVERS
CONSISTS OF :

1. This volume of text.
2. An album of 41 plates.

It is an abreviation of the complete work :

The International Atlas
of
Clouds
and of
States of the Sky.

It is published thanks to the generosity of

**The Paxtot Institute
of Catalonia.**

INTERNATIONAL METEOROLOGICAL COMMITTEE

COMMISSION FOR THE STUDY OF CLOUDS

International Atlas of Clouds and of States of the Sky

ABRIDGED EDITION

FOR THE USE OF OBSERVERS

In memory of our Friend

A. DE QUERVAIN

Member of
the International Commission
for
the Study of Clouds

INTRODUCTION

Since 1922 the International Commission for the Study of Clouds has been engaged in studying the classification of clouds for a new International Atlas. The complete work will appear shortly, and in it will be found a history of the undertaking.

This atlas is only a summary of the complete work, and is intended for the use of observers. The necessity for it was realised by the International Conference of Directors, in order to elucidate the new international cloud code; this is based on the idea of the *state of the sky*, but observers should be able to use it without difficulty for the separate analysis of low, middle, and high clouds. It also explains the place taken in this summary by the essentially practical commentary on the new cloud code. This is also why the extract appears before the complete work, which might seem paradoxical at first sight; but it gives observers a guide to the new code as early as possible.

The second part (aerial observations) and the third part (states of the sky) of the complete atlas have been omitted, as they are not indispensible for the work of ordinary observation. Only 41 of the 174 plates of the complete atlas have been included; they are those which are best adapted to the two colour process, a process which gives clearness and an appearance of relief to the skyscape.

Thus reduced, the work comprises a first part (Clouds) comprising a description of cloud forms, and intsructions for their observation, and a second part (Cloud Codes) which is a practical and detailed commentary on the new code. The plates have been chosen so as to illustrate the first as well as the second part, and they are referred to in the text as much as possible; but they have been arranged in the order of the code, for the atlas should be used principally in its application to the code. It is hoped that the atlas has been produced in such a way that observers may find this application easy; thus the type used for

(1) MM. J. Bjerknes, Keil, and Wehrle were chosen by the Commission for Synoptic Weather Information to revise the work of the International Commission for the Study of Clouds.

each description of the code has been chosen to facilitate references; the detailed definition is in heavy type, the explanatory remarks in ordinary type, the remainder in small type, including those parts which explain how to avoid confusion between different cloud forms. It has been thought best to avoid all considerations of synoptic meteorology in the text, the observer being supposed to ignore the general weather situation; at the same time the observer should not be entirely deprived of the assistance that he might obtain from the connection between the state of the sky and the evolution of disturbances; he will therefore find at the end of the second part a diagram showing where the different forms of low, middle, and high clouds are found in relation to a disturbance.

This atlas is the result of the collective work of the Cloud Commission. Its form, the work of Messrs. BERGERON and WEHRLE, has been decided upon by the Office National Météorologique of France; the English and German translations are due to Messrs. CAVE and KEIL respectively.

It is produced at a remarkably low price, and International Meteorology owes this to the generosity of a member of the Cloud Commission, M. Rafel PATXOT, the Maecenas of the Science of Clouds; we are already indebted to him for the very interesting study of clouds made by la Fundaciô Concepcio Rabell i Cibils; he is almost entirely responsible for the publication of the complete atlas, a work which has been of so great a benefit to this summary.

PARIS, February 25 th. 1930.

E. DELCAMBRE,

President of the International Commission for the Study of Clouds.

PART I

CLOUDS

I. — TABLE OF CLOUD CLASSIFICATION

At nearly all levels clouds may appear under the following forms :

a) *Isolated,* heap clouds with vertical development during their formation, and a spreading out when they are dissolving.

b) *Sheet clouds which are divided up into filaments, scales, or rounded masses,* and which are often stable or in process of disintegration.

c) *More or less* continuous cloud sheets, often in process of formation or growth.

CLASSIFICATION INTO FAMILIES AND GENERA

Family A : HIGH CLOUDS

(Mean lower level 6.000 m.) [1]

Form b } 1. **Genus Cirrus.**
2. **Genus Cirrocumulus.**

Form c } 3. **Genus Cirrostratus.**

Family B : MIDDLE CLOUDS

(Mean upper level 6.000 m.
mean lower level 2.000 m.)

Form a
Form b } 4. **Genus Altocumulus.** [2]

Form c } 5. **Genus Altostratus.**

Family C : LOW CLOUDS

(Mean upper level 2.000 m.
Mean lower level close to the ground)

Form a
Form b } 6. **Genus Stratocumulus.** [2]

Form c } 7. **Genus Stratus.**
8. **Genus Nimbostratus.**

Family D : CLOUDS WITH VERTICAL DEVELOPMENT

(Mean upper level that of the cirrus,
Mean lower level 500 m.)

Form a } 9. **Genus Cumulus.**
10. **Genus Cumulonimbus.**

(1) It should be noted that the heights given are for temperate latitudes, and refer, not to sea level, but to the general level of the land in the region. It should be noted that in certain cases there may be large departures from the given mean heights, especially as regards cirrus, which may be found as low as 3000 metres in temperate latitudes, and in polar regions even almost as low as the surface.

(2) Most altocumulus and stratocumulus clouds come under category *b*; but the varieties cumuliformis and particularly castellatus belong to category *a*.

II. — DEFINITIONS AND DESCRIPTIONS OF THE FORMS OF CLOUDS

I. — CIRRUS (Ci.)

A. — Definition

Detached clouds of delicate and fibrous appearance, without shading, generally white in colour, often of a silky appearance (Pl. 33 to 37).

Cirrus appears in *the most varied forms,* such as isolated tufts (Pl. 35), lines drawn across a blue sky (Pl. 33), branching feather-like plumes, curved lines ending in tufts, &c; they are often arranged in bands which cross the sky like meridian lines, and which, owing to the effect of perspective, converge to a point on the horizon, or to two opposite points (cirrostratus and cirrocumulus often take part in the formation of these bands).

B. — Explanatory Remarks

Cirrus clouds are always composed of ice crystals, and their transparent character is due to the state of division of the crystals.

As a rule these clouds cross the sun's disc without dimming its light. But when they are exceptionally thick they may veil its light and obliterate its contour. This would also be the case with patches of altostratus, but cirrus is distinguished by its dazzling whiteness and silky edges.

Halos[1] are rather rare in cirrus.

Sometimes isolated wisps of snow are seen against the blue sky, and resemble cirrus; they are of a less pure white and less silky than cirrus (Pl. 31); wisps of rain (Pl. 22) are definitely grey, and a rainbow, should one be visible, shows their nature at once, for this cannot be produced in cirrus.

Before sunrise and after sunset, cirrus is sometimes coloured bright yellow or red. These clouds are lit up long before other clouds and fade out much later; sometimes after sunset they become grey. At all hours of the day cirrus near the horizon is often of a yellowish colour,

(1) Cf. p. 6, note (1).

this is due to distance and to the great thickness of air traversed by the rays of light.

Cirrus, being in general more or less inclined to the horizontal, tends less than other clouds to become parallel to the horizon, under the effect of perspective, as the horizon is approached; often on the contrary it seems to converge to a point on the horizon (Pl. 37).

C. — Species

Amongst the more remarkable forms one may note :

1° **Cirrus filosus** (Pl. 33).

More or less straight or irregularly curved filaments (neither tufts nor little points) and without any of the parts being fused together.

2° **Cirrus uncinus** (Pl. 36 and 37).

Cirrus in the shape of a *comma*, the upper part ending in a little *tuft* or *point*.

3° **Cirrus densus** (Pl. 35).

Cirrus clouds with such thickness that without care an observer might mistake them for middle or low clouds.

4° **Cirrus nothus** (Hybrid cirrus) (Pl. 34).

Cirrus proceeding from a cumulonimbus and composed of the debris of the upper frozen parts of these clouds.

D. — Varieties

Ordinary cirrus may appear in many very different forms. One may particularly note the forms *floccus* and *vertebratus* which are really aspects of the varieties *cumuliformis* and *undulatus radiatus* respectively (cf. p. 20).

II. — CIRROCUMULUS (Cicu.)

A. — Definition

A cirriform layer or patch composed of small white flakes or of very small globular masses, without shadows (Pl. 41), *which are arranged in groups or lines, or more often in ripples resembling those of the sand on the sea shore* (Pl. 37).

B. — Explanatory remarks

In general cirrocumulus represents a degraded state of cirrus and cirrostratus both of which may change into it (Pl. 41). In this case the changing patches often retain some fibrous structure in places.

Real cirrocumulus is uncommon. It must not be confused with small altocumulus patches on the edges of altocumulus sheets (Pl. 41). There are in fact all states of transition between cirrocumulus and altocumulus proper; this is only to be expected as the process of formation is the same. In the absence of any other criterion the term cirrocumulus should only be used when:

1° There is evident connection with cirrus or cirrostratus (Pl. 41).

2° The cloud observed, results from a change in cirrus or cirrostratus (Pl. 41).

3° The cloud observed, shows some of the characteristics of ice crystal clouds which will be found enumerated under cirrus (p. 3).

Clear rifts are often seen in a sheet of cirrocumulus.

III. — CIRROSTRATUS (Cist.)

A. — Definition

A thin whitish veil (Pl. 37 to 40) which does not blur the outlines of the sun or moon, but gives rise to halos. Sometimes it is quite diffuse and merely gives the sky a milky look (Pl. 37); sometimes it more or less distinctly shows a fibrous structure with disordered filaments (Pl. 39).

B. — Explanatory Remarks

A sheet of cirrostratus which is very extensive, though in places it may be interrupted by rifts, nearly always ends by covering the whole sky. The border of the sheet may be straight edged and clear-cut (Pl. 40) but more often it is ragged or cut up (Pl. 37).

During the day, when the sun is sufficiently high above the horizon, the sheet is never thick enough to mask the shadows of objects on the ground.

A milky veil of fog is distinguished from a veil of cirrostratus of a similar appearance by the halo phenomena [1] which the sun or the moon nearly always produce in a layer of cirrostratus.

What has been said above of the transparent character and colours of cirrus is true to a great extent of cirrostratus.

C. — Species

Cirrostratus has two principal aspects which correspond to the two following species :

1° Cirrostratus nebulosus.

A very uniform light nebulous veil, sometimes very thin and hardly visible, sometimes relatively dense (Pl. 40), but always without definite details and with halo phenomena.

2° Cirrostratus filosus.

A white fibrous veil, where the strands are more or less definite, often resembling a sheet of cirrus densus from which indeed it may originate (Pl. 38 and 39).

IV. — ALTOCUMULUS (Acu.)

A. — Definition

A layer, or patches composed of lamina or rather flattened globular masses (Pl. 19 to 30), the smallest elements of the regularly arranged layer being fairly small and thin, with or without shading. These elements are arranged in groups, in lines, or waves, following one or two directions (Pl. 19 and 24), and are sometimes so close together that their edges join (Pl. 19, 28, and 30).

The thin and semitransparent edges of the elements often show *irisations* which are rather characteristic of this class of cloud.

From the definition it follows that altocumulus comprises the sub-genera :

1° Altocumulus translucidus (Pl. 19 and 24).

Altocumulus formed of elements whose colour — from dazzling

(I) The following are the principal halo phenomena : — I. A circle of 22° radius round the sun or moon ; this is roughly the angle subtended by the hand placed at right angles to the arm when the latter is extended ; this halo is sometimes, but rarely, accompanied by one of 46° radius. 2. Parhelia, Paraselenae (mock suns or mock moons) luminous patches, often showing prismatic colours, a little over 22° from the sun or moon and at the same elevation. 3. A luminous column (sun pillar) placed vertically above and sometimes below, the sun.
Often only small fragments of these appearances are visible but they are none the less characteristic of high clouds.

white to dark grey — and whose thickness vary much from one example to another, or even in the same layer; the elements are more or less regularly arranged and distinct. In the definition of the elements it is the variation in the transparency of the layer, variable from one point to another, that plays the essential part. There appears in the interstices either the blue of the sky, or at least a marked lightening of the layer of cloud due to a thinning out.

2° Altocumulus opacus (Pl. 28).

Altocumulus sheet which is continuous, at least over the greater part of the layer, and consisting of dark and more or less irregular elements, in the definition of which transparency does not play a great part, owing to the thickness and density of the layer; but the elements show in real relief on the lower surface of the cloud sheet.

B. — Explanatory Remarks

The limits within which altocumulus is met with are very wide.

At the greatest heights, altocumulus, made up of small elements (Pl. 21), resembles cirrocumulus; altocumulus however is distinguished by not possessing any of the following characters of cirrocumulus :

1° Connection with cirrus or cirrostratus.

2° An evolution from cirrus or cirrostratus.

3° Properties due to physical structure (ice crystals) enumerated under cirrus.

At lower levels, where altocumulus may be derived from a spreading out of the tops of cumulus clouds (Pl. 25), it may easily be mistaken for stratocumulus; the convention is that the cloud is altocumulus if the smallest, well defined, and regularly arranged elements which are observed in the layer (leaving out the detached elements which are generally seen on the edges) are not greater than ten solar diameters in their smallest diameters.

When the edge or a thin semitransparent patch of altocumulus passes in front of the sun or moon a *corona* appears close up to them; this is a coloured ring with red outside and blue inside; the colours may be repeated more than once. This phenomenon is infrequent in the case of cirrocumulus and only the higher forms of stratocumulus can show it.

Irisation, mentioned above, is a phenomenon of the same type as the corona; it is a sure mark of altocumulus as distinguished from cirrocumulus or stratocumulus.

Altocumulus clouds often appear at different levels at one and the same time (Pl. 21). Often too they are associated with other types of cloud (Pl. 25).

The atmosphere is often hazy just below altocumulus clouds.

When the elements of a sheet of altocumulus fuse together and make a continuous layer altostratus or nimbostratus is the result (Pl. 28). On the other hand a sheet of altostratus can change into altocumulus. It may happen that these two aspects of a cloud sheet may alternate with each other during the whole course of a day. It is also not rare to have a layer of altocumulus coexisting with a veil resembling altostratus at a height very little less than the altocumulus (Pl. 27), (altocumulus duplicatus, de Quervain).

It is interesting to note that one may often observe filiform descending trails to which the name *virga* has been given (Pl. 22).

C. — Species

Among the most remarkable kinds one may note :

Altocumulus cumulogenitus (Cumulo-stratus, de Quervain).

This is an altocumulus cloud formed by the spreading out of the tops of cumulus, the lower parts of the cumulus clouds having melted away ; the layer has the appearance of altocumulus opacus in the first stages of its growth (Pl. 25).

D. — Varieties

An important variety of altocumulus should be noted namely *altocumulus cumuliformis* (cf. p. 20) which has two different aspects :

Altocumulus floccus (Pl. 30).

Tufts resembling small cumulus clouds without a base, and more or less ragged.

Altocumulus castellatus (Pl. 29).

Cumuliform masses with more or less vertical development, arranged in a line, and resting on a common horizontal base, which gives the cloud a crenellated appearance.

The hoods or veils which form above a cumulus by the uplift of a damp layer, and which may be pierced by the tops of the cumulus (Pl. 2), are considered as a detail of cumulus, and denoted by the term *pileus* attached to the name cumulus; but in reality they are aberrant forms of altocumulus translucidus. Moreover similar clouds, independent of cumulus, can be formed by the same process by the effect of a rising current caused by a mountain or any obstacle. They are then named altocumulus, and they are classed, on account of their form with the variety *lenticularis* (cf., p. 20).

V. — ALTOSTRATUS (Ast.)

A. — Definition

Striated or fibrous veil, more or less grey or bluish in colour (Pl. 17 and 18). This cloud is like thick cirrostratus but without halo phenomena; the sun or moon shows vaguely, with a faint gleam, as though through ground glass. Sometimes the sheet is thin (Pl. 17), with forms intermediate with cirrostratus (cirrostratus translucidus). Sometimes it is very thick and dark (Pl. 18), (altostratus opacus), sometimes even completely hiding the sun or moon. In this case differences of thickness may cause relatively light patches between very dark parts; but the surface never shows real relief, and the striated or fibrous structure is always seen in places in the body of the cloud (Pl. 18).

Every form is observed between high altostratus and cirrostratus on the one hand, and low altostratus and nimbostratus on the other.

Rain or snow may fall from altostratus (altostratus precipitans), but when the rain is heavy the cloud layer will have grown thicker and lower, becoming nimbostratus; but heavy snow may fall from a layer that is definitely altostratus.

From the definition of altostratus it follows that there are three sub-classes :

1° Altostratus translucidus (Pl. 17).

A sheet of altostratus resembling thick cirrostratus ; the sun and the moon show as through ground glass.

2° Altostratus opacus (Pl. 18).

An opaque layer of altostratus of variable thickness which may entirely hide the sun, at any rate in parts, but showing a fibrous structure in some parts.

3° Altostratus precipitans.

A layer of opaque altostratus which has not yet lost its fibrous character, and from which there are light falls of rain or snow, either continuous or intermittent. This precipitation may not reach the ground in which case it forms *virga.*

B. — Explanatory remarks

The limits between which altostratus may be met with are fairly wide (about 5000 to 2000 metres).

A sheet of high altostratus is distinguished from a rather similar sheet of cirrostratus by the convention that halo phenomena are not seen in altostratus, nor are the shadows of objects on the ground visible.

A sheet of low altostratus may be distinguished from a somewhat similar sheet of nimbostratus by the following characters : nimbostratus is of a much darker and more uniform grey, and shows nowhere any whitish gleam or fibrous structure ; one cannot definitely see the limit of its undersurface which has a " soft " look, due to the rain, which may not reach the ground.

The convention is also made that nimbostratus always hides the sun and moon in every part of it, while altostratus only hides them in places behind its darker portions, but they reappear through the lighter parts (Pl. 18).

Careful observation may often detect *virga* hanging from altostratus, and these may even reach the ground causing slight rain. If the sheet still has the character of altostratus it will then be called altostratus precipitans, but not if it has become nimbostratus.

A sheet of altostratus, even if it has rifts in places, has a general fibrous character (Pl. 18). A cloud layer, even a continuous one, which has no fibrous structure, and in which rounded cloud masses may be seen is classed as altocumulus (Pl. 28) or nimbostratus according to circumstances.

Altostratus may result from a transformation of a sheet of altocumulus, and on the other hand altostratus may often break up into altocumulus.

C. — Species

There are many varieties; some of them may be distinguished by adding one of the general qualifying adjectives to the name of the subspecies (e. g. altostratus opacus, undulatus, &c. cf. p. 20).

VI. — STRATOCUMULUS (Stcu.)

A. — Definitions

A layer or patches composed of lamina or globular masses; the smallest of the regularly arranged elements are fairly large; they are soft and grey, with darker parts (Pl. 7 to 10).

These elements are arranged in groups, in lines, or in waves, aligned in one or in two directions. Very often the rolls are so close that their edges join together; when they cover the whole sky, as on the continent, especially in winter, they have a wavy appearance.

From the definition it follows that stratocumulus comprises two kinds :

1° Stratocumulus translucidus (Pl. 9).

A not very thick layer; in the interstices between its elements either the blue sky appears, or at any rate there are much lighter parts of the cloud sheet, which here is thinned out on its upper surface.

2° Stratocumulus opacus (Pl. 10).

A very thick layer made up of a continuous sheet of large, dark, rounded masses; their shape is seen not by a difference in transparency, but they stand out in real relief from the under surface of the cloud layer.

There are transitional forms between stratocumulus and altocumulus on the one hand and between stratocumulus and stratus on the other (Pl. 10).

B. — Explanatory Remarks

The difference between stratocumulus and altocumulus is given under the latter (page 7).

It should also be noted that the cloud sheet called altocumulus by an observer at a small height would appear as stratocumulus to an observer at a sufficient height.

It often happens that stratocumulus is not associated with any clouds of the second or third families; but it fairly often coexists with clouds of the fourth family (Pl. 13 and 14).

The elements of thick stratocumulus (stratocumulus opacus) often tend to fuse together completely, and the layer can, in certain cases, change into nimbostratus. The cloud is called nimbostratus when the cloud elements of stratocumulus have completely disappeared and when, owing to the trails of falling precipitation, the lower surface has no longer a clear cut boundary.

Stratocumulus can change into stratus (Pl. 10), and vice versa. The stratus, being lower, the elements appear very large and very soft, so that the structure of regularly arranged globular masses and waves disappears as far as the observer can see. The cloud will be called stratocumulus as long as the structure remains visible (Pl. 10).

C. — Species

1° Stratocumulus vesperalis (Pl. 7).

This name is given to flat elongated clouds which are often seen to form about sunset as the final product of the diurnal changes of cumulus.

2° Stratocumulus cumulogenitus (Pl. 8).

Stratocumulus formed by the spreading out of the tops of cumulus clouds, which latter have disappeared; the layer in the early stages of its formation looks like stratocumulus opacus.

D. — Varieties

The cloud called *roll cumulus* in England and Germany is designated *stratocumulus undulatus* (cf., p. 20); its wave system is in one direction only. It must not be confused with flat cumulus clouds ranged in line. Stratocumulus often has a *mammatus* (festooned) character (cf., p. 20) that is to say there is a high relief on the lower surface where pendant rounded masses or corrugations are observed, and at times these look as though they would become detached from the cloud. Care must be taken not to confuse this cloud with some kinds of altostratus opacus whose under surface may appear to be slightly corrugated; the latter is distinguished by its fibrous structure.

VII. — STRATUS (St.)

A. — Definition

A uniform layer of cloud, resembling fog, but not resting on the ground (Pl. 11).

When this very low layer is broken up into irregular shreds it is designated fractostratus (Frst).

B. — Explanatory Remarks

A veil of true stratus generally gives the sky a hazy appearance which is very characteristic, but which in certain cases may cause confusion with nimbostratus. When there is precipitation the difference is manifest : nimbostratus gives continuous rain (sometimes snow), precipitation composed of drops which may be small and sparse, or else large (at least some of then) and close together, while stratus only gives a drizzle, that is to say small drops very close together.

When there is no precipitation a dark and uniform layer of stratus can easily be mistaken for nimbostratus. The lower surface of nimbostratus however has always a soft appearance (widespread trailing precipitation, " virga ") ; it is quite uniform and it is not possible to make out definite detail ; stratus on the other hand has a "dryer" appearance, and however uniform it may be it shows some contrasts and some lighter transparent parts, that is, places less dark where the cloud is thinner, corresponding to the interstices between the rolls and globular masses of stratocumulus, but considerably larger, while nimbostratus seems only to be feebly illuminated, and as though lit up from within.

Stratus is often a local cloud, and when it breaks up the blue sky is seen.

Fractostratus sometimes originates from the breaking up of a layer of stratus (Pl. 11), sometimes it forms independently and develops till it forms a layer below nimbostratus, which latter may be seen in the interstices (Pl. 12).

A layer of fractostratus may be distinguished from nimbostratus by its darker appearance, and by being broken up into cloud elements. If these elements have a cumuliform appearance in places the cloud layer is called fractocumulus and not fractostratus [1].

(1) Such fractostratus or fractocumulus clouds may be called nimbus (cf. p. 14, note 1).

VIII. — NIMBOSTRATUS [1] (Nbst.)

A. — Definition

A low, amorphous, and rainy layer, of a dark grey colour and nearly uniform; feebly illuminated seemingly from inside. When it gives precipitation it is in the form of continuous rain or snow.

But precipitation alone is not a sufficient criterion to distinguish the cloud which should be called nimbostratus even when no rain or snow falls from it.

There is often precipitation which does not reach the ground; in this case the base of the cloud is always diffuse and looks " wet " on account of the general trailing precipitation, *virga,* so that it is not possible to determine the limit of its lower surface.

B. — Explanatory Remarks

The usual evolution is as follows : a layer of altostratus grows thicker and lower until it becomes a layer of nimbostratus. Beneath the latter there is generally a progressive development of very low ragged clouds, isolated at first, then fusing together into an almost continuous layer, in the interstices of which however the nimbostratus can generally be seen. These very low clouds are called fractocumulus or fractostratus according as to whether they appear more or less cumuliform or stratiform [1]

Generally the rain only falls after the formation of these very low clouds, which are then hidden by the precipitation or may even melt away under its action. The vertical visibility then becomes very bad.

In certain cases the precipitation may precede the formation of fractocumulus or fractostratus, or it may happen that these clouds do not form at all.

Rather rarely a sheet of nimbostratus may form by an evolution from a stratocumulus.

(1) The introduction of nimbostratus is considered indispensable by the President of the International Committee for the study of clouds, on account of criticisms made on the definition of nimbus in the Provisional Atlas. This modification has not yet been able to be submitted for the approval of the International Meteorological Committee. It is therefore only introduced here subject to its being ultimately approved.

The following are the reasons for this new definition : — the definition of nimbus in the Atlas of 1910 led to some confusion ; in fact, in different countries the name nimbus was given *a*) to a low amorphous rainy, layer, originating directly by change from a descending layer of altostratus. *b*) to very low, dark, ragged clouds isolated at first though not later, which form very often below altostratus or under the rainy layer described above, *a*).

In the present Altas it was intended to give the cloud (*a*) the new name of Nimbostratus, which is a better name than nimbus for a continuous layer which is formed by evolution from altostratus. As to clouds (*b*) they are classed as fractocumulus or fractostratus (according as they are more or less cumuliform) from which they are not distinguished either in shape, or in their mode of formation (turbulence); when however they look dark in colour, owing to special lighting (e. g. the presence of a higher cloud sheet) and so have a very different colour from ordinary fractocumulus or fractostratus, they may if thought necessary, be called nimbus.

IX. — CUMULUS (Cu.)

A. — Definition

Thick clouds with vertical development, the upper surface is dome shaped and exhibits protuberances, while the base is nearly horizontal (Pl. 1 to 3, and 12-13).

When the cloud is opposite to the sun the surfaces normal to the observer are brighter than the edges of the protuberances. When the light comes from the side, the clouds exhibit strong contrasts of light and shade; against the sun, on the other hand, they look dark with a bright edge (Pl. 1).

True cumulus is definitely limited above and below, its surface often appears hard and clear cut (Pl. 2). But one may also observe a cloud resembling ragged cumulus in which the different parts show constant change (Pl. 1). This cloud is designated fractocumulus (frcu).

B. — Explanatory Remarks

Typical cumulus (Pl. 1) develops on days of clear skies, and is due to the currents of diurnal convection; it appears in the morning, grows, and then more or less dissolves again towards the evening.

Cumulus whose base is generally of a grey colour, has a uniform structure, that is to say it is composed of rounded parts right up to its summit, with no fibrous structure (Pl. 2). Even when highly developed, cumulus can only produce light precipitation.

Cumulus, when it reaches the altocumulus level, is sometimes capped with a light, diffuse, and white veil of more or less lenticular shape, with a delicate striated or flaky structure on its edges; it is generally shaped like a bow which may cover several domes of the cumulus, and finally be pierced by them (Pl. 2). This cloud which does not constitute a species is given the name of *pileus*, a cap or hood.

The clouds which form below altostratus or nimbostratus and which can develop into a complete layer, through whose interstices the altostratus or nimbostratus is generally seen, are usually fractostratus; but if they have a cumuliform appearance (Pl. 17) they should be classed as fractocumulus (cf. p. 13 and p. 14 note 1). They rarely have this appearance during or soon after rain; on the other hand it is frequent at the beginning of the formation of the low cloud, and when it breaks up.

C. — Species

Among the more remarkable species one may note :

1° **Cumulus humilis** (Pl. 1).

Cumulus with little vertical development, and seemingly flattened. These clouds are generally seen in fine weather.

2° **Cumulus congestus** (Pl. 2).

Very distended and swollen cumulus, whose domes have a cauliflower appearance.

X. — CUMULONIMBUS (Cunb.)

A. — Definition

Heavy masses of cloud, with great vertical development, whose cumuliform summits rise in the form of mountains or towers, the upper parts having a fibrous texture and often spreading out in the shape of an anvil (Pl. 4 to 6, and 16).

The base resembles nimbostratus, and one generally notices *virga.* This base (Pl. 16) has often a layer of very low ragged clouds below it (fractostratus, fractocumulus, cf. p. 13 and 14 note 1).

Cumulonimbus clouds generally produce showers of rain or snow (Pl. 6), and sometimes of hail or soft hail, and often thunderstorms as well.

If the whole of the cloud cannot be seen the fall of a real shower is enough to characterise the cloud as a cumulonimbus.

B. — Explanatory Remarks

Even if a cumulonimbus were not distinguished by its shape from a strongly developed cumulus its essential character is evident in the difference of structure of its upper parts, when these are visible (fibrous structure and cumuliform structure). Masses of cumulus however heavy they may be, and however great their vertical development, should never be classed as cumulonimbus unless the whole or a part of their tops is transformed (Pl. 5) or is in process of transformation (Pl. 4) into a cirrus mass.

Although the upper cirriform parts of a cumulonimbus may take on very varied shapes, yet in certain cases they spread out into the form of an anvil (Pl. 5). To this interesting variety the name *incus* is given.

In certain types of cumulonimbus, which are especially common in spring in moderately high latitudes, the fibrous structure extends to nearly the whole cloud mass (Pl. 16), so that the cumuliform parts almost wholly disappear ; the cloud is reduced to a mass of cirrus and of *virga*.

The veil cloud *pileus* is seen with cumulonimbus clouds as with cumulus.

When a cumulonimbus covers nearly all the sky the base alone is visible, and resembles nimbostratus (Pl. 16), with or without fractostratus or fractocumulus below[1]. The difference between the base of a cumulonimbus and a nimbostratus is often rather difficult to make out. If the cloud mass does not cover all the sky, or if even small portions of the upper parts of the cumulonimbus appear, the difference is evident. If not it can only be made out if the preceeding evolution of the clouds has been followed, or if precipitation occurs ; its character is violent and intermittent (showers) in the case of cumulonimbus, as opposed to the relatively gentle and continuous precipitation of a nimbostratus.

The front of a thunder cloud of great extent is sometimes accompanied by a roll cloud of a dark colour in the shape of an arch, of a frayed out appearance, and circumscribing a part of the sky of a lighter grey. This cloud is named *arcus* and is nothing more or less than a particular case of fractocumulus or fractostratus[1] (Pl. 16).

Fairly often *mammatus* structure appears in cumulonimbus, either at the base, or on the lower surface of the lateral parts of the anvil.

When a layer of menacing cloud covers the sky and *virga* and *mammatus* structure are both seen it is a sure sign that the cloud is the base of a cumulonimbus, even in the absence of all other signs.

Cumulonimbus is a real *factory of clouds* ; it is responsible in great measure for the clouds in the rear of disturbances. By the spreading out of the more or less high parts and the melting away of the underlying parts, cumulonimbus can produce either more or less thick sheets of altocumulus or stratocumulus (spreading out of the cumuliform parts —Pl. 8), or dense cirrus (spreading out of the cirriform parts—Pl. 34).

C. — Species

Amongst the remarkable species may be noted :

1° Cumulonimbus calvus (Pl. 4).

Cumulonimbus characterised by the thunderstorm or the shower

(1) Cf. p. 13 and p. 14, note (1).

that it causes, or by *virga*, but in which at first no cirriform parts can be made out. Nevertheless the freezing of the upper parts has already begun; the tops are beginning to lose their cumulus structure, that is to say their rounded outlines and clear cut contours; the hard and "cauliflower" swellings soon become confused and melt away so that nothing can be seen in the white mass but more or less vertical fibres (Pl. 4). The freezing, accompanied by the change into a fibrous structure, often goes on very rapidly.

2° **Cumulonimbus capillatus** (Pl. 5).

Cumulonimbus which displays distinct cirriform parts, having sometimes (Pl. 5), but not always, the shape of an anvil.

III. — INSTRUCTIONS FOR THE OBSERVATION OF CLOUDS

I. — DETERMINATION OF THE CLOUD FORM, VARIETIES AND CASUAL DETAILS

At the time of each observation it is essential to determine the *family* to which the clouds belong (high cloud, middle cloud, low cloud, or cloud with vertical development).

The observer will then specify and enter in the observation book :

1° The *genus* of the cloud described by the international code number used in the Atlas. It must be remembered that *typical forms are relatively rare* ; it is generally forms more or less intermediate that are observed; consequently one must determine the typical form which the cloud observed most closely resembles, making use of the plates and descriptions in the Atlas.

2° The *species* (the particular kind of cloud belonging to the genus already determined) making use of the definitions, the illustrations, and the names given in the Atlas under the cloud in question.

3° The *variety* (that is one of the particular forms common to different genera) using the definitions and abreviations given below.

4° The *casual details,* which do not characterise either species or varieties, according to the definitions given below.

Finally if an observed cloud closely resembles a cloud reproduced in the Atlas the observer would do well to note the number of the plate in question.

If the cloud is in process of evolution the observer should note both the present and the preceeding forms.

A) PRINCIPAL VARIETIES

The chief varieties common to different genera are as follows :

1° Fumulus *(Fum.)*.

At all levels, from cirrus to stratus, a very thin veil may form, so delicate that it may be almost invisible. These veils seem to be most frequent on hot days, and in low latitudes. Occasionally they may be observed to thicken rapidly, forming clouds easily visible, especially cirrus and cumulus. The clouds thus produced seem unstable however, and usually melt away soon after their formation.

Cirrus fumulus must not be confused with cirrostratus nebulosus. The latter is much more stable and does not show the phenomenon of the formation and subsequent rapid disappearance of cirrus clouds.

2° **Lenticularis** *(Lent.)*.

Clouds of an ovoid shape, with clean cut edges, and sometimes irisations, especially common on days of föhn, sirocco, and mistral winds. This form exists at all levels from cirrostratus to stratus. See plate 20 for altocumulus lenticularis.

3° **Cumuliformis** *(Cuf.)*.

The rounded form resembling cumulus which the upper parts of other clouds may sometimes assume. This may be seen at all levels from cirrus to stratus. See plate 29 for altocumulus castellatus, and plate 30 for altocumulus floccus.

4° **Mammatus** *(Mam.)*.

This description is given to all clouds whose lower surfaces form pockets or festoons. This form is found especially in stratocumulus and in cumulonimbus, either at the base, or even more often on the lower surface of anvil projections. It is also found, though rarely, in cirrus clouds, probably when they have originated in the anvil of a dispersing cumulonimbus.

5° **Undulatus** *(Und.)*.

This term is applied to clouds composed of elongated and parallel elements, like waves of the sea. It is well to note the orientation of these lines or waves. When there is an appearance of two distinct systems, as when the cloud is divided up into rounded masses by undulations in two directions, the observer will note the orientation of the two systems. Observation should be made on lines as near the zenith as possible to avoid errors due to perspective. See plates 19 and 23 for altocumulus undulatus.

6° Radiatus *(Rad.)*.

This term is applied to clouds in parallel bands (polar bands), which owing to perspective seem to converge to a point on the horizon, or to two opposite points if the bands cross the whole sky. The point is called the Radiant point, or Vanishing point, and its position on the horizon should be noted (N, NNE... &c.). See plate 24 for altocumulus radiatus. The point on the horizon should be noted where the bands, if prolonged, would converge, if they do not actually reach the horizon.

B) CASUAL VARIETIES

The chief casual varieties are the following :

1° *Virga*, wisps or falling trails of precipitation; applied principally to altocumulus and altostratus. See plate 20 for altocumulus virga.

2° *Pileus*, a cap or hood ; applied principally to cumulus or cumulonimbus. See plate 2 for cumulus pileus.

3° *Incus*, anvil ; applied to cumulonimbus. See plate 5 for cumulonimbus incus.

4° *Arcus*, arch cloud ; applied to cumulonimbus. See plate 16 for cumulonimbus arcus.

C) SUPPLEMENTARY CHARACTERS

1° In the case of *veil clouds* (cirrostratus or altostratus) one may also note :

a) The *thickness* of the veil, using the following scale :

0. Very thin and irregular.
1. Thin but regular.
2. Moderately thick.
3. Thick.
4. Very thick, and of a dark colour.

b) The *direction* in which the veil is *thickest*.

2° The *optical phenomena* (halo, corona, irisation &c.); it should be stated whether the phenomena are *evanescent* or *persistent*.

3° An observation of the state of the sky should also contain an estimate of the *cloudiness* ; both the *total cloud amount*, that is the amount of the sky, in tenth parts, covered by all the clouds, and the *partial cloud amounts* for each genus of cloud, that is the amount of the sky, in tenths, that would be covered by each genus of cloud if it were alone in the sky.

In estimating the cloud amount it is best to ignore the part of the sky close to the horizon, for there the clouds seem to be packed together owing to perspective, so that the cloud amount seems greater than it really is. When the amount of cloud is large it is easier to estimate the amount of the sky that is free from cloud, and from this, by subtraction, to obtain the cloud amount.

In estimating partial cloud amounts a difficulty occurs when there are superposed sheets or patches of cloud. But it is generally feasible to wait till the patches of low cloud, whose apparent movement is usually rapid, have moved away, so that the previously hidden parts of the high layer can be seen.

II. — DIRECTION OF MOTION AND VELOCITY OF THE CLOUDS

Measurements of the above complete the observation of the sky.

A) MEASUREMENT OF THE DIRECTION OF MOTION AND OF THE ANGULAR VELOCITY

a) An observer with a comb, or other type of nephoscope, can easily determine the direction of motion of the clouds and their angular velocity.

Apparatus for upper air observations may be used as a nephoscope, as long as the clouds are not too high, and if the magnification is not too great (otherwise the cloud would appear too soft so that it would be difficult to fix on a definite point). A definite point of the cloud is followed exactly as a pilot balloon is followed during an ascent.

b) If the observer has no special instrument he can proceed as follows, by making use of any vertical line (mast, lightning conductor &c.) : a definite point in the cloud is chosen, not too far from the zenith ; the observer takes up his position at such a distance from the mast that his line of sight to the cloud point passes through the top of the mast, and in such a direction that the cloud point seems to move down or up the length of the mast ; the direction of the straight line from the feet of the observer to the base of the mast gives the direction of cloud motion in the first case, and the reverse of the direction in the second case.

If for any reason it is found that one cannot get into such a position that the cloud seems to move down or up the length of the mast, one can place oneself in any position at such a distance from the mast that the line of sight from a definite cloud point passes through the top of the mast. As the cloud moves the observer moves too, keeping the point of the cloud " on " with the top of the mast. The ground is marked with the heel at the beginning and end of the operation to mark the *direction* of cloud motion ; the *sense* of the direction is obtained by remembering that the observer has moved in an *opposite* direction to the movement of the cloud.

As regards the apparent or angular movement an observer without a nephoscope can only roughly estimate the motion on a simple scale : slow (displacement almost imperceptible), moderate (displacement quite appreciable) and fast [1].

B) DETERMINATION OF THE TRUE MOTION OF A CLOUD

To obtain the true motion one must multiply the angular motion by the height, but the height cannot generally be determined at all accurately.

At a station where pilot balloons are available however the height can be determined from the time between the liberation of the balloon and its disappearance in the cloud layer.

In mountainous countries the height of a cloud layer can often be roughly determined by using the natural datum marks afforded by the mountains.

In general the estimation of the height of a cloud from its genus or species may lead to very large errors.

III. — THE NECESSITY OF NOTING THE STATE OF THE SKY AS A WHOLE AND OF FOLLOWING THE EVOLUTION OF THE CLOUDS

A) THE NECESSITY OF NOTING THE STATE OF THE SKY AS A WHOLE

It is evident from the specifications of the cloud code (second part, p. 27) that to describe the sky at the station at a particular time logically and completely it is not enough to know the genera or even the species of the clouds present ; for example altocumulus appears in seven

(1) When there are several layers it is of interest to note their apparent relative velocities.

specifications of the code, and cirrus in nine. In reality each specification of the code, as the explanations show, is not so much a dry enumeration of the genera or species of clouds in the sky, as a general indication of the structure, the organization, and the evolution of the cloud complex which makes up the state of the sky. Some specifications only refer to the general structure; for example $C_M = 9$ is a thundery sky; everyone knows that in thundery conditions degenerate cloud forms are met with which are very difficult to classify, while the thundery look of the whole sky is apparent immediately and without any doubt.

Each specification of the code corresponds to a state of the sky lower, middle, or high. The observer should have at his fingers ends the commentaries accompanying the definitions; he should consider as a whole the lower, middle, or high clouds there described, and try to make a considered judgement of the observed sky as a whole, so that he can directly apply to it a number of the code.

The detailed analysis of the individual clouds should follow and not precede this recognition of the state of the sky as a whole. If the observer gets used to this course he will find in a short time that the different states of the sky, lower, middle and high, corresponding with the code, will seem just as " live " as the typical cloud forms, and it will be just as easy to identify a state of the sky as the form of a cloud.

B) NECESSITY OF FOLLOWING THE EVOLUTION OF THE SKY

The aspect of the sky is continually changing, and many transitional forms exist between the different types of cloud described in the Atlas. It is relatively rare for the observer to see typical clouds of one genus which float past, or persist in the sky for any considerable time; in most cases he will find that he has difficulties at the time of observation if he has not taken the trouble to watch the sky since the last observation. If however he has taken this precaution he will often be able to refer a confusing state of the sky, or a particular cloud, to a previous state which was typical and easy to identify. Moreover most of the specifications of the cloud code take into account the evolution of the clouds. A single isolated observation is insufficient.

As regards evolution, the recognition of the state of the sky as a whole, recommended in the previous paragraph, is better than the identification of clouds considered by themselves, for as a matter of fact the evolution of the state of the sky can be followed indefinitely at one station, while the evolution of a cloud, if, as is usual, it is a " migrant " can only be observed during the relatively short time that it takes to cross the sky.

SECOND PART

CLOUD CODE

I. — LOWER CLOUDS C_L

0.— No lower clouds.

1.— Cumulus of fine weather (L1 — Pl. 1).

This is observed under different aspects : (1) in a state of formation, in general in the morning (2) Completely formed in general in the middle of the day, with definite horizontal bases, the air being more or less calm. These are *a*) with rounded tops but without " cauliflower " heads, or *b*) flat and " deflated " (3). Completely formed but broken up by the wind ; in this case they remain separate and are white in colour. The cumulus clouds of fine weather usually have a marked diurnal period, growing until the middle of the afternoon, and decreasing later, both as to amount of cloud in the sky and in their vertical development. The photograph L1 (Pl. 1) corresponds to the second class (variety *b* with some traces of *a*); individual clouds of the third class may also be noticed.

These cumulus clouds are only found away from disturbances. When the veil of cirrostratus, which fringes the front of a disturbance, begins to cover the sky, the cumulus of fine weather changes from type 2 *a*) to type 2 *b*) and finally disappears entirely.

The fractocumulus of fine weather noticed above as type 3, and coded as $C_L = 1$, must not be confused with the fractocumulus of bad weather [1] which is coded $C_L = 6$, or $C_L = 9$ (see p. 30 to 31). The former are detached white clouds in a blue sky, and remain detached ; the latter are found in the central part of a disturbance or in its rear ; in the first case ($C_L = 6$) they form under a grey sheet of altostratus or of nimbostratus ; in the second case ($C_L = 2$), in a sky crowded with clouds at all altitudes, they form under the bases of cumulonimbus or very large cumulus clouds or in the spaces between these. In both cases they are dark, receiving little light, and generally become very numerous, while the fractocumulus clouds of fine weather show white on a blue sky, and remain detached.

2.—Cumulus, heavy and swelling, without anvil top (L2*a* — Pl. 2 and L2*b* — Pl. 3).

There are two types of these :

1° In calm air and especially on hot days with a thundery tendency they form heavy masses with horizontal bases, and very great vertical development (L2*a* — Pl. 2) they are sometimes in the form of towers, sometimes of complex heaps with " cauliflower " formation. They are then often capped with hoods, *pileus* (L2*a'* — Pl. 2).

2° In strong winds in the rear of disturbancies they also form towering masses, with great vertical development but tossed about and broken up. (L2*b* — Pl. 3.)

These cumulus clouds, especially those of the second class, are often associated with thick cirrus ($C_H = 3$) and with extensions of stratocumulus and altocumulus ($C_M = 6$). None of these cumulus clouds should show ice crystal clouds (hybrid cirrus) at their tops ; this would mean that they had reached the cumulonimbus stage which would entail code $C_L = 3$.

3.—Cumulonimbus (L3*a* — Pl. 4, L3*b* — Pl. 5, and L3*c* — Pl. 6).

Cumulus clouds of great vertical development, with the tops composed of ice crystal clouds. Sometimes the nascent ice crystal cloud is merely mingled with the " cauliflower " tops, where a fibrous structure appears and the clear-cut outlines fray out (L3*a* — Pl. 4), sometimes the completely formed ice crystal clouds crown the cumulus with a definite plume of cirrus of a shape more or less that of an anvil (L3*b* — Pl. 5). Sometimes especially in the spring and in high latitudes the ice crystal formation involves nearly the whole cloud even to the base (L3*c* — Pl. 6). At the end of the growth of a cumulonimbus the lower cumuliform part of the cloud often tends to disappear leaving only the upper or cirrus part.

The photographs L3*a* (Pl. 4), and L3*b* (Pl. 5), represent fairly distant cumulonimbus clouds seen in elevation; Photograph L3*c* (Pl. 6) shows a less simple aspect; the cloud is not seen as a whole by the observer, nor in its normal proportions; this aspect of a cumulonimbus approaching the observer is however frequent; the anvil, reaching nearly to the zenith, begins to overshadow the observer. In this case, a *mammatus* structure will often be seen on the lower surface of the anvil projection.

Like the heavy and swelling cumulus, cumulonimbus is formed

either in calms, especially on hot thundery days, or in a strong wind in the rear of disturbances.

Cumulonimbus is a regular *factory of clouds* (L4*b* — Pl. 5). By extension at various levels it often produces either cirrus masses by an extension of the ice crystal parts, or masses of altocumulus or stratocumulus by an extension of the cumuliform parts, and these may end by becoming detached from the parent cloud. Thus cumulonimbus, $C_L = 3$, may coexist with cloud sheets that should be coded $C_H = 3$ or $C_M = 6$ (cf. M6 — Pl. 25).

At the end of the evolution of cumulonimbus, $C_L = 3$ should only be coded when the cumuliform parts are still visible, otherwise $C_H = 3$ should be coded. The photograph H3*a* (Pl. 34) represents cumulonimbus anvils which have lost their cumuliform parts.

When cumulonimbus nears the zenith, and its base, with low dark clouds under it, often in the form of a roller or an arch, has covered all or nearly all the sky, code $C_L = 3$ should be replaced by $C_L = 9$. Photograph L9*b* (Pl. 16) shows such a sky, where the black roller shows very clearly. Photograph L3*c* (Pl. 6) shows a cumulonimbus when the anvil has nearly reached the zenith; it is a view intermediate between that shown in photographs L3*a* (Pl. 4) and L3*b* (Pl. 5) where a fairly distant cumulonimbus is clearly seen in elevation, and that shown in photograph L9*b* (Pl. 16) where the base of a cumulonimbus in the zenith covers all the sky.

4.—Stratocumulus formed by the flattening of cumulus clouds (L4*a* — Pl. 7, and L4*b* — Pl. 8).

Cumulus tops may settle down (L4*a* — Pl. 7) and the bases may spread out; this is a frequent end to the progressive changes of the cumulus of fine weather. On the other hand (L4*b* — Pl. 8), the bases may melt away and the tops may spread out; this is a common phenomenon in the rear of a disturbance after squalls or showers. Very opaque sheets or a layer of stratocumulus may be formed in this way, often showing a festooned formation *(mammatus)* in places. At the end of the process the clouds thus formed may thin out.

The first case is *stratocumulus vesperalis*, the second *stratocumulus cumulogenitus*.

5.—Layer of stratus or stratocumulus (L5*a* — Pl. 9, L5*b* — Pl. 10, and L5*c* — Pl. 11).

Clouds usually forming a single layer, fairly regular and not very dark or menacing; they have a certain stability. The stratocumulus has often semitransparent parts, or even clear spaces between the elements of the cloud (L5*a* — Pl. 9).

Photograph L5*a* (Pl. 9) shows a layer of stratocumulus, L5*c* (Pl. 11) a layer of stratus, and L5*b* (Pl. 10) a transitional state, stratocumulus tending towards stratus.

These cloud formations are common on the continent, especially in winter, and are found outside the regions of disturbances, or on their extreme lateral edges.

The layer of stratocumulus may often be broken up ; code $C_L = 5$ is only used for those sheets of stratocumulus that are not formed from cumulus ; otherwise they are coded $C_L = 4$. The observer may be in doubt between $C_L = 5$ and $C_M = 3$; the code $C_L = 5$ is only used when the stratocumulus is fairly low and rather like stratus (large and rather diffuse tessellations or waves) ; if it is clearly high up and related to altocumulus it is coded $C_M = 3$.

6.—Low broken up clouds of bad weather [1] (L6 — Pl. 12).

The following is the ordinary course of formation of these clouds : when a veil of altostratus becomes lower, and tends to turn into nimbostratus, it usually has below it a gradually increasing layer of fractocumulus or fractostratus ; these clouds are isolated at first. Plate M1 shows the beginning of the fractocumulus forming below a typical altostratus. They ultimately fuse together into a continuous sheet ; but through interstices the veil of relatively light higher cloud may be seen (L6 — Pl. 12). The continuous rain does not usually occur until after the formation of the fractostratus or fractocumulus, which is then hidden by the precipitation, or may even disappear under its influence.

This type is found in the middle of a typical disturbance.

For the difference between the fractocumulus of bad weather and that of fine weather see the explanation under $C_L = 2$.

7.—Cumulus of fine weather and stratocumulus (L7 — Pl. 13).

Cumulus clouds may form below an already existing sheet of stratocumulus, which they do not penetrate.

Before the formation of the cumulus the layer of stratocumulus would have been coded $C_L = 5$ or $C_M = 3$ according to its height.

As in photograph L7 (Pl. 13) there should not be any continuous merging of the tops of the cumulus clouds with the layer above them ; if that were the case it would show that the layer was formed from the extension of the tops of the cumulus, and it should in that case be coded $C_L = 4$, or if the layer were not too low $C_M = 7$, and at the same time the cumulus should be coded $C_L = 1$.

If the cumulus penetrates the stratocumulus layer it must be coded $C_L = 8$.

8.—Heavy or swelling cumulus, or cumulonimbus, and stratocumulus (L8 — Pl. 14).

Heavy or swelling cumulus, or cumulonimbus may form below an already existing sheet of stratocumulus, and some of the cumulus or cumulonimbus may penetrate the layer.

(1) These clouds represent one of the forms of Fractostratus or Fractocumulus, as described in the International Atlas for 1930, wich may, if considered necessary, be called nimbus, cf. p. 14, note (1).

This is a formation analagous to $C_L = 7$ except that the vertical development of the convection clouds is more marked, so that in the absence of the upper layer they would be coded $C_L = 2$ or $C_L = 3$. In photograph L8 (Pl. 14) it is clearly shown that some of the cumulus penetrates the stratocumulus layer.

9.—Heavy or swelling cumulus (or cumulonimbus) and low ragged clouds of bad weather [1] (L9*a* — Pl. 15 and L9*b* — Pl. 16).

When a heavy or swelling cumulus or a cumulonimbus nears the zenith it can hide all, or nearly all the sky with its base ; the latter may somewhat resemble nimbostratus, but is distinguished from it by its previous history, or by the broken character of the rain, which, in the case in question, falls to the ground in showers, or, if it does not reach the ground, is seen falling from the lower surface of the cloud in wisps of rain or snow called virga (cf. M4c — Pl. 22). Generally below the base of such a great cloud there is a greater or smaller amount of lower cloud broken up in the manner of fractocumulus or fractostratus (L9*a* — Pl. 15), these low clouds may sometimes be in the form of a *roller*, or an *arch* (L9*b* — Pl. 16).

It may also happen that low dark clouds of the nature of fractocumulus or fractostratus may grow in amount in a sky covered with heavy or swelling cumulus or with cumulonimbus and may fill the spaces between the main cumulus clouds. In L9*a* (Pl. 15) it should be noticed that the low ragged clouds have already invaded the bases of the cumulus and cumulonimbus clouds.

Only in these two cases should $C_L = 9$ be coded.

1° When a cumulus cloud approaches the zenith and its upper parts are still visible it should be coded $C_L = 2$ or $C_L = 3$ according as to whether it is a heavy and swelling cumulus or a cumulonimbus. L3*c* — Pl. 6 corresponds to such a case intermediate between the clouds of plates 4 (L3*a*) and 5 (L3*b*) and those of photograph L9*b* (Pl. 16).

2° In damp climates it may happen that dark fractocumulus or fractostratus form a layer completely closing up the spaces between the masses of heavy cumulus. In such cases one cannot distinguish individual cumulonimbus clouds under their typical aspect ; but their passage overhead is manifest by a temporary darkening of the sky and by showers. Their presence, thus made known, allows the code $C_L = 9$ to be used and not merely $C_L = 6$.

(1) These clouds represent one of the forms of Fractostratus or Fractocumulus, as described in the International Atlas for 1930, which may if considered necessary be called nimbus cf. p. 14, note (1).

II. — MIDDLE CLOUDS C_M

0.—No middle clouds.

1.—Typical altostratus, thin (M1 — Pl. 17).

A sheet of this cloud resembles thick cirrostratus, from which it is often derived without any break; but halo phenomena, sun pillar, &c., are not seen in altostratus, and the sun appears as though shining through ground glass and does not cast shadows.

This cloud is found in the central part of a typical disturbance.

If there were halo phenomena or if the sun cast shadows one would code $C_H = 5$, $C_H = 6$, $C_H = 7$, or $C_H = 8$ according to circumstances. If the sun were hidden, or liable to be completely hidden by a thick part of the sheet, the code $C_M = 2$ should be used.

In Photograph M1 (Pl. 17) it should be noted that some fractocumulus $C_L = 6$ appears below the altostratus.

2.—Typical altostratus, thick [1] (M2 — Pl. 18).

The sun and moon are completely hidden, at any rate by some parts of the cloud sheet. Typical thick altostratus can be formed either by a thickening of typical thin altostratus, $C_M = 1$ or by the fusing together of the cloudlets in a sheet of altocumulus, $C_M = 7$.

Nimbostratus is derived either by a change from typical thick altostratus, or by the fusing together of the cloudlets in a sheet of stratocumulus ($C_L = 5$).

This type is found in the central parts of a typical disturbance.

In photograph M2 (Pl. 18) the sun still shows vaguely through a relatively thin part of the altostratus, but at G it would be completely hidden by a thicker part of the sheet.

In the case of a transition from altocumulus into altostratus, if no fibrous structure is visible in the layer, and if it shows either in whole or part the structure of altocumulus (ripples, waves, or tesselations) it is coded $C_M = 7$. In the case of a change from stratocumulus to nimbostratus, it will also be coded $C_M = 7$ if the lower surface shows a real relief (waves or tesselations) instead of showing a smooth under surface. Thick typical altostratus and nimbostratus are often accompanied by underlying and very low clouds, which are ragged and dark (fractocumulus or fractostratus); in the gaps one can generally see the altostratus or the nimbostratus of a lighter grey; this case will be coded as $C_L = 6$, or $C_M = 2$. If the lowest clouds form a continuous sheet the observer will not assume what is above, but will code $C_L = 6$, $C_M = /$, $C_H = /$ (middle and upper clouds invisible).

3.—Altocumulus, or high stratocumulus, sheet at one level only (M3 — Pl. 19).

A cloud generally forming a single layer; it is fairly regular, and of uniform thickness, the cloudlets, tesselations or waves, being always

(1) The Nimbostratus of the International Atlas for 1930 will be similarly coded.

separated by clear spaces or lighter gaps ; the cloudlets are neither very large nor very dark. This layer is generally fairly stable, that is to say it changes but slowly. In tropical and subtropical regions, including the Mediterranean, such a layer often forms at the end of the night in calm weather, even at a high altitude, in the absence of any disturbance.

This layer of cloud is sometimes broken up but it is only coded as $C_M = 3$ if the layer does not proceed from the extensions of the tops of cumulus clouds ; in the latter case it is coded as $C_M = 6$. For the difference between $C_M = 3$ and $C_L = 5$ see the explanatory remarks on $C_L = 5$. The layer of altocumulus that should be coded $C_M = 3$ is distinguished from that coded $C_M = 5$ by its stability with no tendency to increase, and by a greater regularity and uniformity.

4.—Altocumulus [1] in small isolated patches, individual clouds often showing signs of evaporation and being more or less lenticular in shape (M4*a* — Pl. 20, M4*b* — Pl. 21 et M4c — Pl. 22).

As regards the smallness of the cloudlets these little patches of altocumulus resemble cirrocumulus, but do not show the characters of clouds formed of ice crystals. Lenticular altocumulus clouds show the most beautiful *irisations* ; when this is the case they are lens shaped, fairly thick, but with little or no shadows and of a pure dazzling white; they are slightly wavy at the edges. (Photograph M4*a* —Pl. 20). Generally (M4*b* — Pl. 21) they are scattered over the sky quite irregularly and are often at different levels ; they are mostly in constant change so that if one looks away from the sky for only a few minutes it is difficult afterwards to identify the clouds previously seen. Individually they are often in process of dissolution, but the amount of cloud over the whole sky does not in general become greater or less.

This type of cloud is common on the extreme lateral edge of disturbances and also in mountainous districts under the influence of the föhn wind.

The characteristics mentioned above are very typical. Nevertheless in cases where the observer is doubtful between $C_M = 4$ and $C_M = 6$ or between $C_M = 4$ and $C_M = 5$ they can be discriminated in the following way: the sheets of altocumulus of $C_M = 4$ are higher and more delicate than those of $C_M = 6$ and they have not the same regular structure as those of $C_M = 5$. Wisps of rain or snow (*virga*) may sometimes be seen falling from the underside of sheets of altocumulus of class $C_M = 3$. In photograph M4*c* (Pl. 22) the lenticular form of an altocumulus sheet is well seen near the horizon; in the same plate a wisp of rain, considerably larger than is usually observed, is seen falling from a sheet near the zenith, but does not reach the ground.

(1) Under this denomination will also be found certain forms of the cirrocumulus described as such in the International Cloud Atlas for 1910.

5.—Altocumulus arranged in more or less parallel bands, or an ordered layer advancing over the sky (M5*a* — Pl. 23 and M5*b* — Pl. 24).

The essential feature of this type is that the sky becomes more and more covered; the process often begins (photograph M5*a* — Pl. 23) with altocumulus in large parallel bands, which often have a roughly lenticular shape. On their edges they may show signs of evaporating, but on the whole the amount of cloud and the thickness of the sheets increases.

Sometimes altocumulus appears from the beginning as a sheet (photograph M5*b* — Pl. 24), often under the form of a vast pavement with more or less rectilinear joints, and the semitransparent parts allow the blue of the sky to be seen. This layer very soon thickens in places, or has another layer, lower and darker, forming beneath it.

This cloud is met with on the lateral side of a typical disturbance or on the front of a weak disturbance.

This type of altocumulus is distinguished from that coded as $C_M = 3$ by the progressive deterioration of the sky, and by the irregular thickness of the layer. Again the proportions of these sheets and their regular structure differentiate them from those coded as $C_M = 4$.

6.—Altocumulus formed by a spreading out of the tops of cumulus (M6 — Pl. 25).

Cumulus clouds of sufficiently great vertical development may undergo an extension of their summits while their bases may gradually "melt away". These sheets of altocumulus are generally fairly thick and opaque at first (M6 — Pl. 25), with rather large elements, dark and soft; later they may thin out and finally have rifts in them, or at any rate semitransparent interstices.

The phenomenon of the extension of cumuliform masses is common in the rear of disturbances, after squalls and showers.

These extensions from the summits of cumulus clouds must not be confused with the ice crystal extensions of a cumulonimbus from which the *anvil* and hybrid cirrus is formed. The anvil, and even the cirrus, detached from the cumulus which produced it, may sometimes have a festooned appearance on the lower surfaces, and thus may have a certain likeness to a sheet of altocumulus. But the altocumulus does not have the whiteness, the silky appearance, nor the fibrous structure of a cirrus anvil.

To discriminate between the altocumulus coded $C_M = 6$ on the one hand, and the altocumulus coded $C_M = 3$, $C_M = 4$, and $C_M = 5$ on the other see the explanations of $C_M = 3$, $C_M = 4$, and $C_M = 5$.

The observer may in certain cases be in doubt between codes $C_M = 6$ and $C_L = 5$. At the end of the evolution of sheets of altocumulus cumulogenitus when they may have thinned out and become semitransparent one would not of course think of coding $C_L = 5$; but at the beginning of the change when the cloud elements are rather large, dark, and soft there are all the transitions between stratocumulus and altocumulus. It is a question of the apparent size of the elements ; the convention is that the cloud in question is altocumulus when the smallest well defined cloud elements observed over the sheet as a whole (and ignoring the separated portions which may exist beyond the edges of the sheet) are not greater than ten diameters of the sun ; in case of doubt it is best to code $C_M = 6$, which allows one to note ($C_L = 2$ or $C_L = 3$) cumulus clouds which are visible at the same time.

7.—Altocumulus associated with altostratus, or altostratus with a partially altocumulus character (M7*a* — Pl. 26, M7*b* — Pl. 27, and M7c — Pl. 28).

Different clouds are comprised in this section :

1° Typical altostratus can lie above sheets of altocumulus which are at a definitely lower level (photograph M7*a* — Pl. 26). This type including typical altostratus can only exist in the central part of a disturbance.

2° A more or less continuous layer of altocumulus can have below it a grey veil of cloud, often hardly visible, lying at a level very little lower ; in places, and for short times it hides the cloudlets of the altocumulus sheet sufficiently to give it in places the appearance of altostratus (photograph M7*b* — Pl. 27). This type is found in the central part of a weak disturbance, or on the lateral edge of the central part of a typical disturbance.

3° In a layer of altocumulus which is growing thicker ($C_M = 5$) the cloudlets may fuse together and the layer may become altostratus, the cloud loses its tessellations, and takes on a fibrous structure; this transitional state is coded $C_M = 7$, as is also the transition from stratocumulus to nimbostratus. It is observed in the same situation as the preceeding cloud.

4° Again altostratus, and even nimbostratus can change progressively into altocumulus, and in this case also the state of transition is coded $C_M = 7$. This phenomenon is common after continuous rain, when the altostratus breaks up, that is to say at the passing away of the end of the central part of a typical disturbance. It is also found in the actual centre when the disturbance is in course of filling up.

5° The opaque cloud sheets with a more or less irregular corrugated structure, too dense and thick for the transparence of the ripples to afford any criterion for their classification, should be coded as $C_M = 7$. The wave structure, though not visible as lighter parts, is nevertheless apparent, and gives a corrugated appearance to the under surface. These sheets were sometimes called " wrinkled " altostratus in the old classification, but they should now be called altocumulus opacus (Photograph M7c—Pl. 28). They are met with in the same situations as classes 2, and 3 above.

These different types have this in common, that they denote a weakening of a disturbance, because they are related either to one which is weakened as a whole, or to a weakened, lateral part of a typical disturbance.

As regards the usually ephemeral change of altostratus, or nimbostratus to altocumulus or stratocumulus, or vice versa (class 3 and 4 above) and the differences between codes $C_M = 2$ and $C_M = 7$, see the notes on $C_M = 2$.

Corrugated altostratus opacus (class 5 above) has two origins ; either it forms *a*) like the cloud sheets of class 3 above, by the uniting of the cloudlets and the general thickening of a layer of altocumulus coded as $C_M = 5$ (or more rarely $C_M = 3$), or *b*) as in class 4 above, it forms from thick altostratus or from nimbostratus ($C_M = 2$) which is in process of dissolution ; but it represents a definitely more stable type than 3 or 4 ; in the case of these latter there is a quick and almost direct change of the altocumulus translucidus into altostratus or vice versa ; in the case of class 5 the intermediate stage of altocumulus opacus can be maintained for some time. The layer of corrugated altocumulus opacus ($C_M = 7$) is distinguished from sheets of altocumulus coded $C_M = 3$ or $C_M = 5$ by its greater thickness which gives it a darker colour and a menacing appearance, and by the very marked irregularity of the relief of the under surface which gives it a characteristic pendant appearance (festoon clouds, pendant corrugations).

8.—**Altocumulus castellatus, or scattered cumuliform tufts** (M8*a* — Pl. 29, and M8*b* — Pl. 30).

The character common to these types of altocumulus is a domed shape, but they have very different aspects.

1° Altocumulus castellatus (M8*a* — Pl. 29) is composed of a series of small cumuliform masses with more or less vertical development, arranged in line and resting on a common horizontal base (reduced sometimes to plain grey wisps); this gives the cloud a crenellated appearance.

2° The second kind (M8*b* — Pl. 30) is observed in scattered tufts, white or grey, but without definite shadow, with the rounded parts very slightly domed ; they resemble very small, more or less broken up cumulus clouds without a base.

These cloud forms are the precursers of thunderstorms, but the first named variety may appear a long way in advance.

9.—Altocumulus [1] **in several sheets at different levels, generally associated with thick fibrous veils of cloud, and a chaotic appearance of the sky** (M9*a* — Pl. 31, and M9*b* — Pl. 32).

This type is very difficult to analyse; it is generally complex, and patches of middle cloud, more or less fragmentary, are seen superposed; they are often badly defined and with soft outlines; there are all the transitional forms between low altocumulus and the fibrous veil (M9*a* — Pl. 31). The sky moreover is covered with clouds of different layers, but as in general there is not a continuous sheet, blue patches may be seen (M9*b* — Pl. 32). Luckily though these middle clouds are very difficult to classify individually the state of the whole sky is very typical; it has (M9*b* — Pl. 32) a *disordered, heavy,* and *stagnant* appearance (calm or light wind).

These appearances are characteristic of the central parts of thundery disturbances.

The clouds of other levels which almost always exist in a thundery sky are principally cirrus densus ($C_H = 3$) and large cumulus ($C_L = 2$) or cumulonimbus ($C_L = 3$).

It would be too complicated, and hardly worth while, to attempt to analyse in detail the individual characters of the altocumulus sheets in a thundery sky. It should be carefully noted that such a sky can be identified by its general appearance as a whole, which is quite typical. There is only one other type which in this respect may be confused with a thundery sky, because it also has a more or less disordered appearance; this is the sky in the rear of a disturbance, but this latter is fairly easily distinguished by its "tossed about" and "windswept" appearance.

(1) Under this name will be found certain forms of the cirrocumulus of the International Cloud Atlas for 1910.

UPPER CLOUDS C_H

0.—No upper clouds.

1.—Cirrus, delicate, not increasing scattered and isolated masses.

This type of isolated cirrus is widely scattered; its amount does not noticeably increase either in time or in any particular direction. It does not collect into sheets or bands, and there is no tendency for the elements to fuse together into masses of cirrostratus. The cirrus clouds whose strands end in an upturned hook or tuft must not be included in this class.

Cirrus clouds of this type are indications of a very distant disturbance, and are found either on its front or on its borders. The place where these clouds are observed is generally not influenced by the disturbance, at any rate for some time.

The chief characteristic of the type $C_H = 1$ is the sparseness of the cirrus. It is only distinguished from the type $C_H = 2$ (see photograph H2 — Pl. 33) by this fact of the cirrus being more sparse.

This type of cirrus is distinguished : 1° from that of class $C_H = 3$ in that it is delicate and does not originate in anvil clouds ; 2° from class $C_H = 4$ in that it does not increase either in time or in any particular direction of the sky, and that it does not belong to the uncinus type (strands with upturned ends) ; 3° from class $C_H = 5$ and class $C_H = 6$ in that it does not increase in time or in any particular direction, that it does not form into sheets or bands, and that it does not show, in any part of the sky, any tendency to pass into cirrostratus.

2.—Cirrus, delicate, not increasing, abundant but not forming a continuous layer (H2 — Pl. 33).

The definition of this type is precisely the same as the preceeding, with the exception that in this type the cirrus is more abundant over the *whole* sky, but without any tendency to increase in any particular direction.

Cirrus of this type is the forerunner of a disturbance either on its front or on its borders.

3.—Cirrus of anvil clouds, usually dense (H3*a*— Pl. 34 and H3*b*—Pl. 35).

In this type the cirrus either actually proceeds from the anvil of a cumulonimbus, in which case one sees cirriform masses attached to the remains of cumulus H3*a* — Pl. 34, or it has probably so originated, being still anvil like in shape, or else remarkable (H3*b* — Pl. 35) for its density and frayed out appearance, generally showing *virga* in places.

Originating in cumulonimbus this type of cirrus is met with either in the rear of typical disturbances or else round thunderstorms.

In the first case the cloud is hybrid cirrus ; in the second it is dense cirrus, in which case the cumulonimbus stage may have ended some time previously. Photograph H3*b* (Pl. 34), shows dense cirrus originating probably in the anvil of a thundercloud, though it has lost the anvil character.

4.—Cirrus, increasing, generally in the form of hooks ending in a point or in a small tuft (H4 — Pl. 36).

This type of cirrus which is often in the form of streaks ending in a little upturned point or in a small tuft, photograph H4 (Pl. 36), increases in amount both in time and in a certain direction. In this direction it reaches the horizon where there is a tendency for the cloud elements to fuse together, but the clouds do not pass into cirrostratus.

This type often occurs on the front of a typical disturbance.

If the cirrus increases in time and in a certain direction, tending also to pass into cirrostratus in this direction it should be coded as $C_H = 6$ or $C_H = 5$ according as to whether the front of the cirrus sheet, formed out of the scattered cirrus is at a greater or lesser altitude than 45° above the horizon.

5.—Cirrus (often in polar bands), or cirrostratus [1] advancing over the sky, but not more than 45° above the horizon (H5 — Pl. 37).

Sheet of fibrous cirrus (H5 — Pl. 37) partly uniting into cirrostratus, especially towards the horizon in the direction where the cirrus strands tend to fuse together ; the cirrus is often in a herring-bone formation, or is in great bands converging more or less to a point on the horizon. In this class is also included a sheet of cirrostratus without any cirrus. In either case the front of the sheet is not more than 45° above the horizon.

This type occurs in the front of a typical disturbance.

The varieties included in this definition are " herring-bone cirrus " and " polar bands ". What is termed the front of the sheet means the front either of the cirrostratus sheet, if this cloud exists without cirrus, or of that part of the sky where the cirrus fibres are close enough to each other to appear welded together, instead of being discrete masses like the cirrus which precedes this sheet.

6.—Cirrus (often in polar bands) or cirrostratus [1] advancing over the sky, and more than 45° above the horizon (H6 — Pl. 38).

The definition of this type is exactly the same as the previous one, with the sole exception that the edge of the sheet is more than 45° above the horizon.

(1) Cirrus and cirrostratus may coexist.

This type is found in the front part of a typical disturbance, a little nearer the centre than the former type.

The same remarks on the limits of the cloud sheet apply as for $C_H = 5$.

Photograph H6 (Pl. 38) represents a case of a veil of rather thin cirrostratus when the fibrous structure is clearly visible. Judging by the position of the sun and the time of day this cirrostratus cloud is certainly at an altitude greater than 45° ; but to code it as $C_H = 6$ we must consider that it does not cover the whole sky, otherwise it would be coded as $C_H = 7$.

7. — Veil of cirrostratus covering the whole sky (H7 — Pl. 39).

There are two cases : 1° Thin, very uniform nebulous veil sometimes hardly visible, sometimes relatively dense, always without definite detail, but always producing halo phenomena (halo, sun pillar) round the sun and moon. 2° A white fibrous sheet (H7 — Pl. 39) with more or less clearly defined fibres, often like a sheet of fibrous cirrus, from which indeed it may be derived.

This type is found in a typical disturbance immediately in front of the central area.

8.—Cirrostratus not increasing, and not covering the whole sky (H8 — Pl. 40).

This is a case of a veil or sheet of cirrostratus (H8 — Pl. 40) reaching the horizon in one direction but leaving a segment of blue sky in the other direction ; this segment of blue sky does not grow smaller. Generally the edge of this sheet is clear cut, and does not tail off into scattered cirrus.

This type is found on the *north* lateral edge of a disturbance, the aspect of the cloud being generally very different from that of the clouds on the *south* lateral edge ; (the terms north and south apply to the most frequent case of a disturbance moving from west to east).

If the segment of blue sky is *diminishing* the cloud must be coded as $C_H = 7$ or $C_H = 6$ according as to whether the elevation of the sheet of cirrostratus is greater or less than 45°.

9.—Cirrocumulus [1] **predominating, associated with a small quantity of cirrus** (H9 — Pl. 41).

An absolutely necessary characteristic of cirrocumulus, according to the new definition of this cloud, is its relationship to cirrus or cirrostratus.

(1) *N. B.* — The definition of this cloud is that of the Atlas of 1930. It should be noted that Cirrocumulus may be present in each of the skies described in sections I. to 8.

This type is found on the front or lateral edges of a weak disturbance.

If the cirrus fibres, or the sheet of cirrostratus merely becomes slightly corrugated in parts (H5 — Pl. 37) the cirrocumulus is neglected. But if the bank of cirrus or of cirrostratus degenerates wholly into cirrocumulus (H9 — Pl. 41) it is coded as $C_H = 9$; this is an important sign of the weakening of a disturbance.

IV. — DISTRIBUTION (AROUND A DISTURBANCE) OF THE SKY AND CLOUDS CORRESPONDING WITH THE DIFFERENT SPECIFICATIONS OF THE CODE

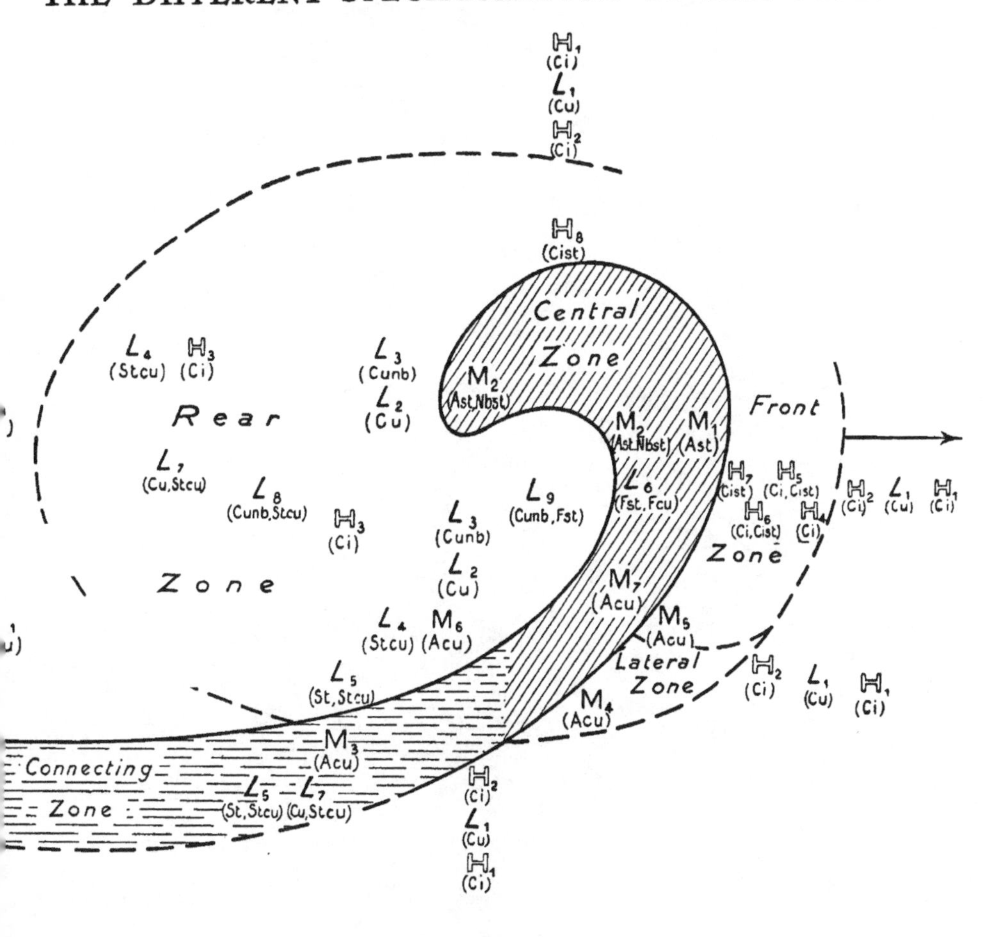

NOTE

I. — This diagram corresponds with a typical disturbance — to be precise, the first of a family — arriving from the West in Western Europe.

II. — It sometimes happens that the rear zone is much more extensive and that it persists for several days over the same region.

III. — Fractocumulus may occur practically anywhere in the rear zone.

IV. — The specifications M8 and M9 correspond respectively with the front or lateral sector and with the central or rear sector of a *thunderstorm* disturbance. They can find no place in this diagram which is that of a *normal* disturbance.

V. — Observers should not take the form of cloud corresponding with the position in this diagram as necessarily determining the form of cloud to be reported.

NOTE

TABLE OF CONTENTS

PART I
CLOUDS

PART II
CLOUDS CODE

II. — MIDDLE CLOUDS C_M

III. — UPPER CLOUDS C_H

IV. — DISTRIBUTION (AROUND A DISTURBANCE) OF THE SKY AND CLOUDS CORRESPONDING WITH THE DIFFERENT SPECIFICATIONS OF THE CODE

Meteorologisch-Magnetisches Observatorium, Potsdam, Sept. 23rd 1916, 15 h., looking SW, altitude 27°.

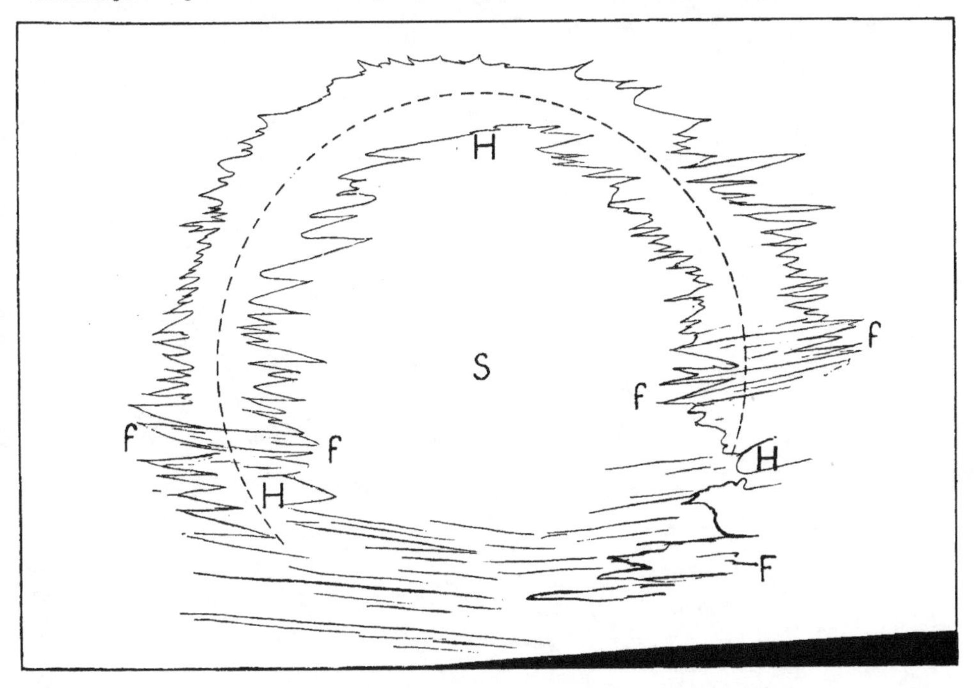

Cirrostratus increasing in amount, and more than 45° above the horizon. — *Code number* **H 6.** — A fibrous sheet of cirrostratus with the sun at **(S)** and a halo at **(HHH)**, practically the whole circle being visible. The fibrous structure **(ff)** of the sheet of cirrostratus is well seen near the halo. Low down, at **(F)**, the cirrus clouds are noticeably thicker; high up on the other hand they are much thinner, and are only revealed by the halo; one can infer from this that the sheet of cirrostratus, though it is higher than 45°, does not cover all the sky.

F. W. Baker.

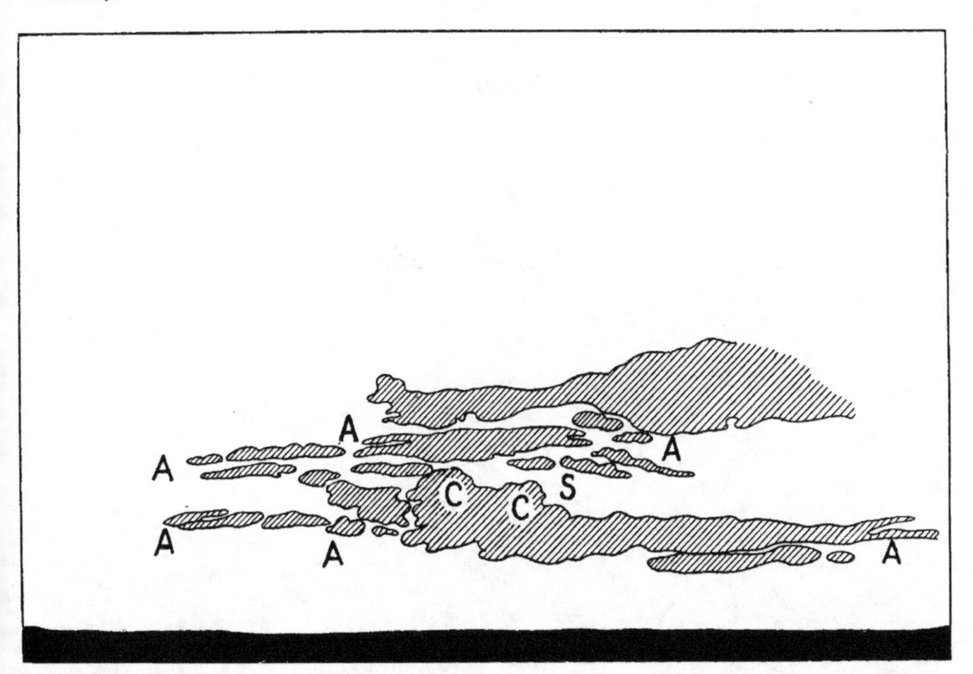

Stratocumulus formed by extension from Cumulus (Stratocumulus vesperalis. — Melting away of the tops and spreading out of the bases). — *Code number* **L 4.** — The time is that of the end of the diurnal formation of cumulus. The clouds, in the course of disappearing, are almost completely flattened, and resemble lines of stratus **(AA)**, dark against the setting sun **(S)**. At **(CC)** some traces of rounded cumulus can still be seen.

Meteorologisch-Magnetisches Observatorium, Potsdam, May 24th 1929, 13 h. 55, looking NE.

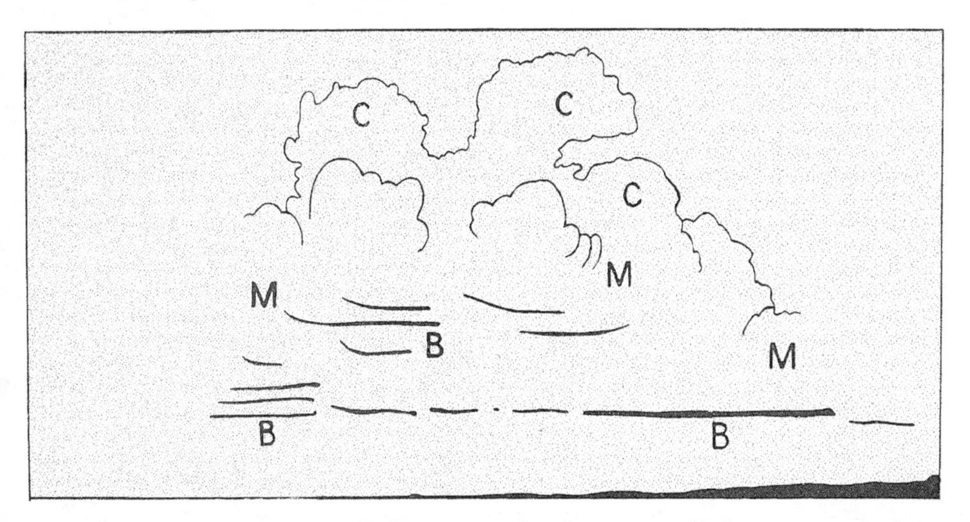

Heavy Cumulus without anvil (Cu. congestus). — Code number **L 2.** — The very definite horizontal bases **(BB)** and the vertical and symetrical development of the cloud masses show that the air is calm. Notice the complex character of the clouds which are formed by the fusion of more or less distinct heavy masses **(MM)**. These have been built up by successive growths, and the tops have the characteristic " cauliflower " appearance **(CC)**; the surfaces look " hard, " and the contours show clear cut against the sky.

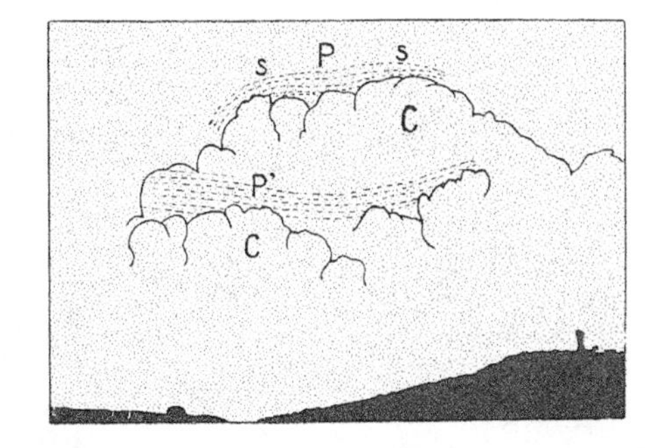

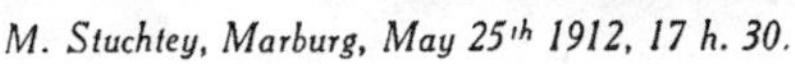

M. Stuchtey, Marburg, May 25th 1912, 17 h. 30.

Cumulus pileus.— Code number **L 2.** — This photograph. taken from a mountain, gives a closer view of the " cauliflower " heads. Light, soft veils (pileus) are seen in places **(PP')** above the tops. The veil **(P)** has a striated structure very different from the rounded masses of the cumulus, and looks as though it were lifted up by the cloud heads.

Office National Météorologique, Paris, Aug. 12th 1925, 13 h. 03, looking S, altitude 27°.

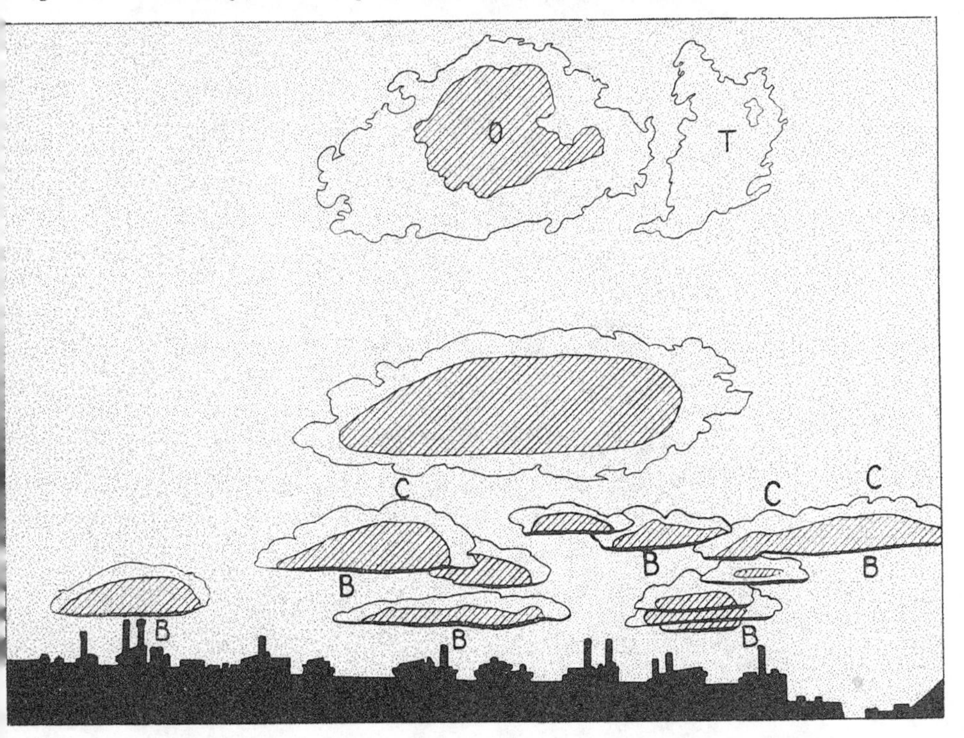

Cumulus of fine weather (Cumulus humilis). — *Code number* **L 1.** — The clouds are scattered and have a flat and deflated appearance even at the diurnal maximum of cloud development in the early afternoon. Their horizontal extension is greater than the vertical, as can be seen directly in the clouds near the horizon, and indirectly in the clouds near the zenith; these have small areas of shadow **(O)** or are even transparent **(T)** which shows that their thickness is not great. At **(CC)** the clouds are slightly rounded. Near the horizon the bases **(BB)** are clearly shown. The cloud **(T)** which has no horizontal base and whose edges are ragged is a fractocumulus.

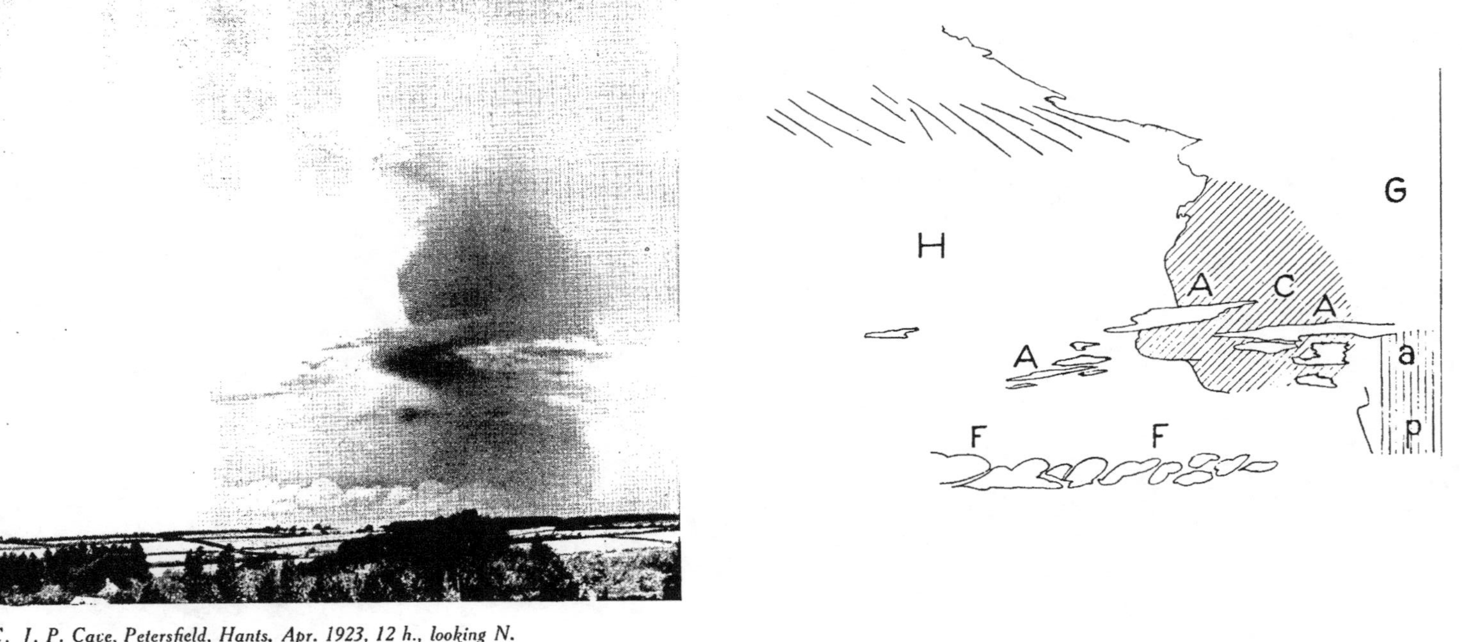

C. J. P. Cave, Petersfield, Hants, Apr. 1923, 12 h., looking N.

Cumulonimbus. — *Code number* **L 3.** — The cloud is nearing the zenith. The different parts of the cloud are not very clearly shown, as happens frequently in similar cases. There is a shower of snow or hail as far down as **(p)**, which is at the level at which the snow melts and falls as rain, or else where the shower evaporates without reaching the ground. At **(C)** is a very dark cumuliform part of the cloud. At **(E)** is the anvil. At **(G)** above the falling snow it is evident that the freezing point comes down very low. At **(H)** is cirrostratus, probably the anvil of another cumulonimbus. At **(AA)** is fractostratus of the base, or rather degraded stratocumulus formed by extension from the main cloud. At **(FF)** is cumulus and fractocumulus.

F. W. Baker, Farnborough, Hants, April 14th 1923, 11 h 30, looking SW.

Cumulonimbus (Cumulonimbus incus). — *Code number* **L 3.** — The typical anvil **(EE)** is completely formed ; it is seen in elevation because the cumulonimbus is fairly far off, and it has the typical shape. The cirrus mass shows a striated structure **(SS)** right down to the main cloud mass, in spite of its great thickness, as shown by the shadow it casts ; the edges of the anvil are frayed out, showing a structure very different from the rounded cumulus forms at **(CC)**.

G. A. Clarke, Aberdeen, July 1905, looking SSE, altitude 5°

Layer of Stratus. — *Code number* **L 5.** — The cloud is very low and appears to be very uniform, for the observer is too near to distinguish any wave structure at the high elevation at which the cloud is seen. The cloud descends onto the hill at **(C)** and hides the top. Shreds of cloud (fractostratus) sweep along the hillside at **(F)**.

G. A. Clarke, Aberdeen, Feb. 27th 1907, 14 h. 30, looking SW, altitude 25°.

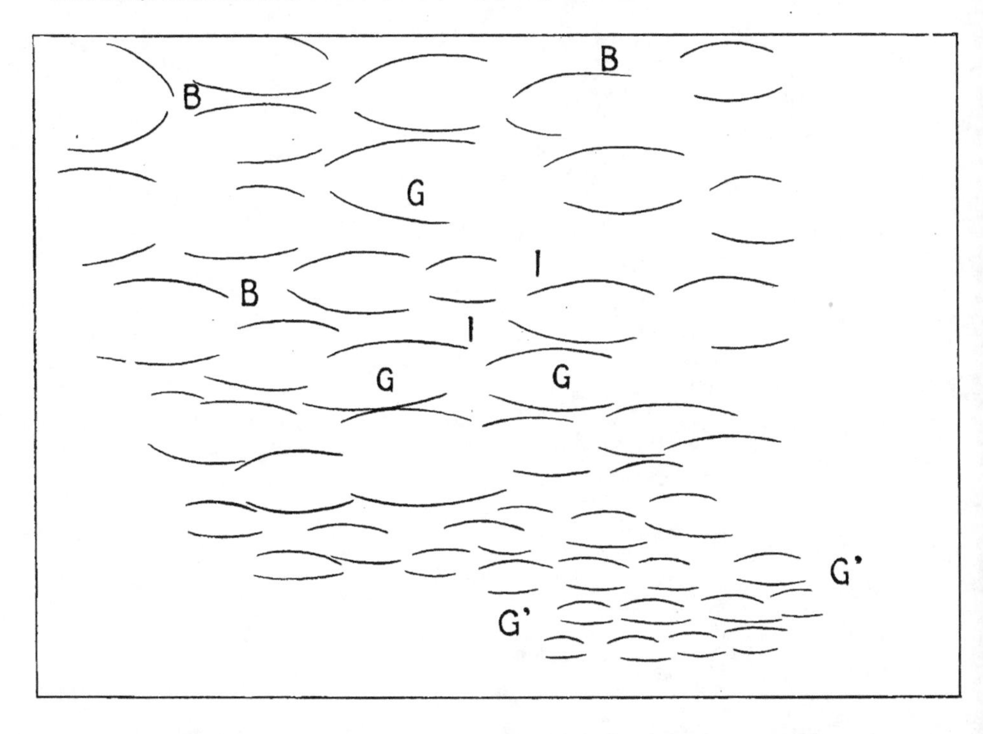

Layer of Stratocumulus (Stratocumulus translucidus). — *Code number* **L 5.** — The cloud elements **(GG)** have a globular shape, intermediate between a sphere and a flat slab; they form a fairly regular layer. They have dark shadows and therefore are fairly thick, but in the interstices the layer is much thinner and very light; sometines there are even some rifts and blue sky appears **(BB)**. On the horizon **(G'G')** perspective produces an appearance of rollers due to the allignment of the cloud elements, which are thus shown to have a rather regular arrangement.

Fundació Concepció Rabell, Barcelona, Nov. 23rd 1925, 12 h. 40 to 13 h.

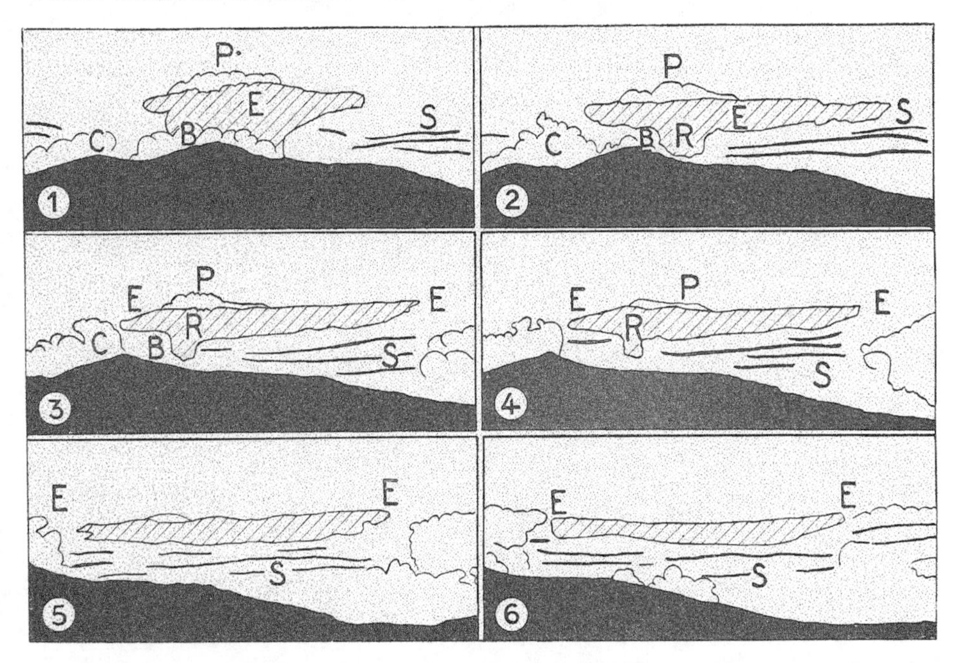

Stratocumulus formed by the spreading out of Cumulus (Stcu. cumulogenitus. — Melting away of the base, and spreading out of the top). — Code number **L 4.** — The time from the beginning to the end of the series is 20 minutes. 1. It is clear that the top of the cloud that is growing out is of the rounded cumulus type **(B)**, without any cirriform parts. The spreading out has begun at **(E)** and the head of the cumulus has penetrated the extension at **(P)**. 2. **(P)** has developed a little, but **(B)** is decreasing and **(E)**, which is increasing in extent, is beginning to separate off, so that the extreme base of the cloud **(R)** is now seen. 3. **(P)** grows smaller, **(B)** has completely settled down and is detached at **(R)**, while **(E)** is still developing in extent. 4. **(P)** has completely settled down, **(R)** is melting away, **(E)** is completely independent. 5 and 6. There is no longer any trace of **(P)** or **(R)**, while **(E)** is fully formed; notice the pendant cloud particles on the lower surface. On all the photographs other bands of stratocumulus may be seen in the distance which are being drawn out into stratus; they probably originated in the same way.

G. A. Clarke, Aberdeen, May 9th 1917, 15 h. 20.

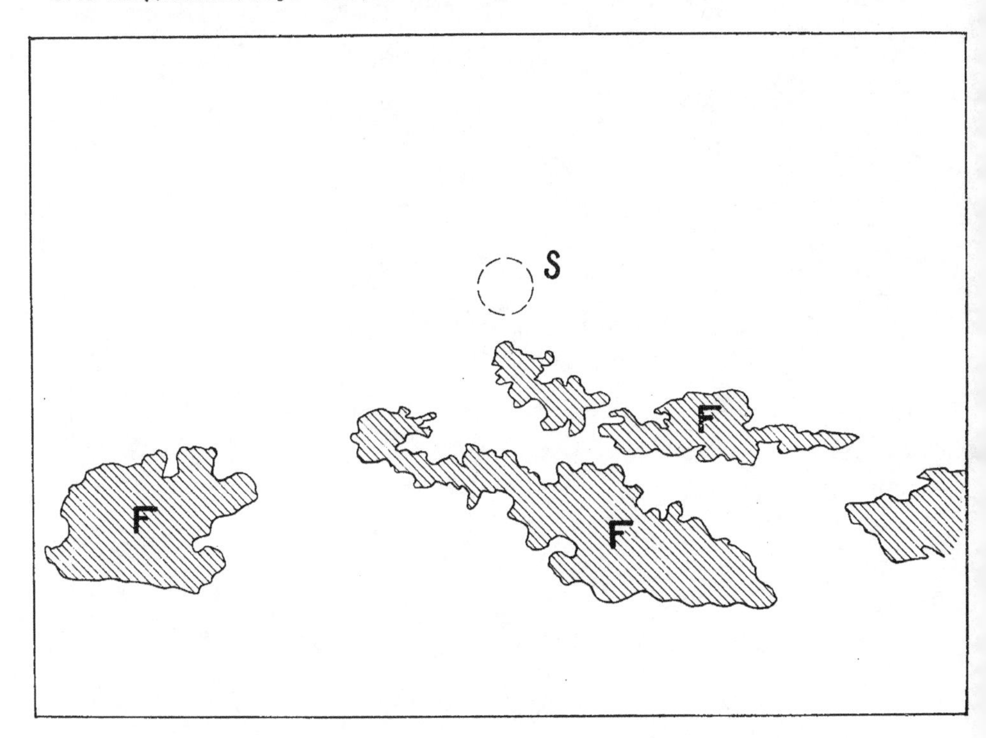

Typical thin Altostratus (Altostratus translucidus). — *Code number* **M 1.** — The altostratus is thin and semi-transparent. The sun still shows fairly definitely **(S)**, and the cloud is not therefore nimbostratus; but one can no longer distinguish the sun's outline, and there are no halo phenomena: the cloud is therefore not cirrostratus. At **(FF)** are small fractostratus clouds with a slight tendency towards cumulus; the veil of altostratus being between them and the sun, they appear very dark.

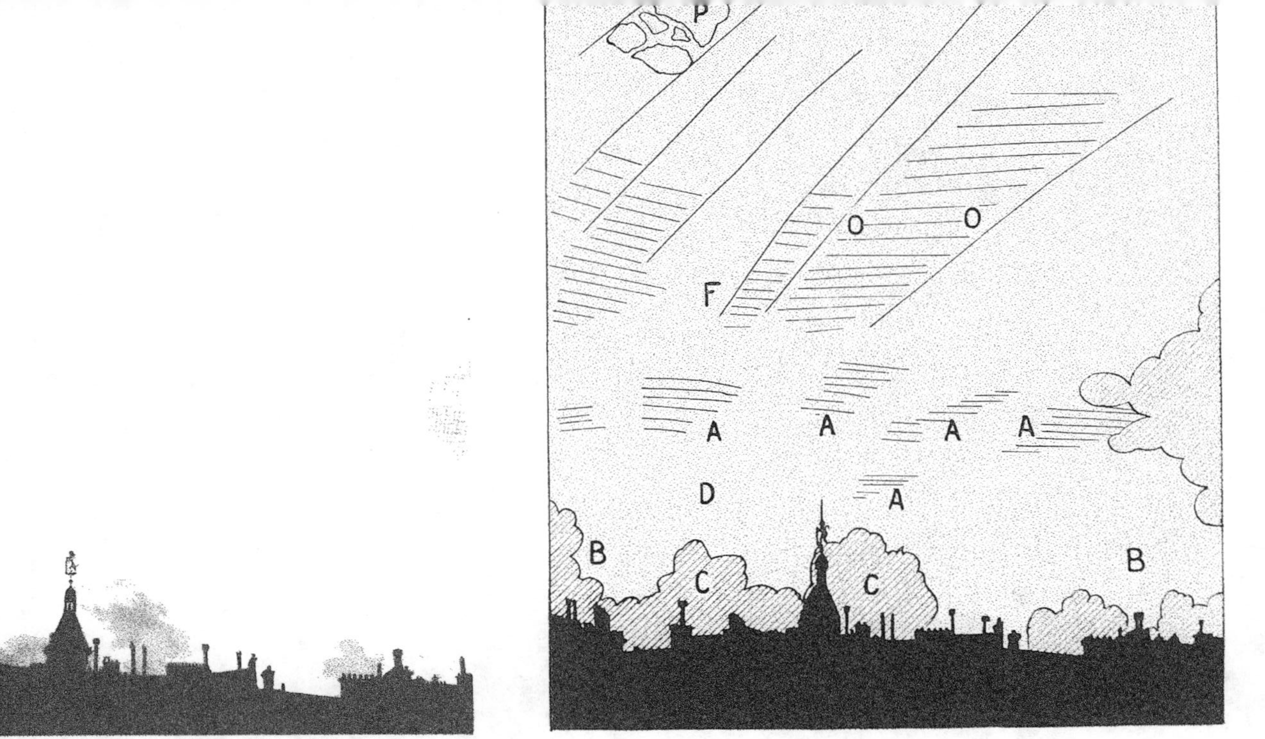

Office National Météorologique, Paris, Aug. 6th 1925, 9 h. 45, looking S, altitude 25°.

Altocumulus in a definitely organized layer spreading over the sky (Altocumulus translucidus). — *Code number* **M 5.**
— A vast pavement with more or less rectilinear joints, the cloud elements being fairly thin lamina ; these are arranged in lines **(FF)** in the direction of view ; a second system of waves at right angles to the first can be seen at **(OO)**. At **(D)** is a uniform layer of cloud that one might take for cirrostratus but the structure at the edge of the sheet at **(AA)** shows that we are dealing with the same layer of altocumulus. There are cumulus clouds **(CC)** slightly developed and half hidden in a very hazy atmosphere (especially at **AA**) ; this bad visibility is frequent in weak disturbances.

Office National Météorologique, Paris, Sept. 12th 1916, 8 h. looking ENE.

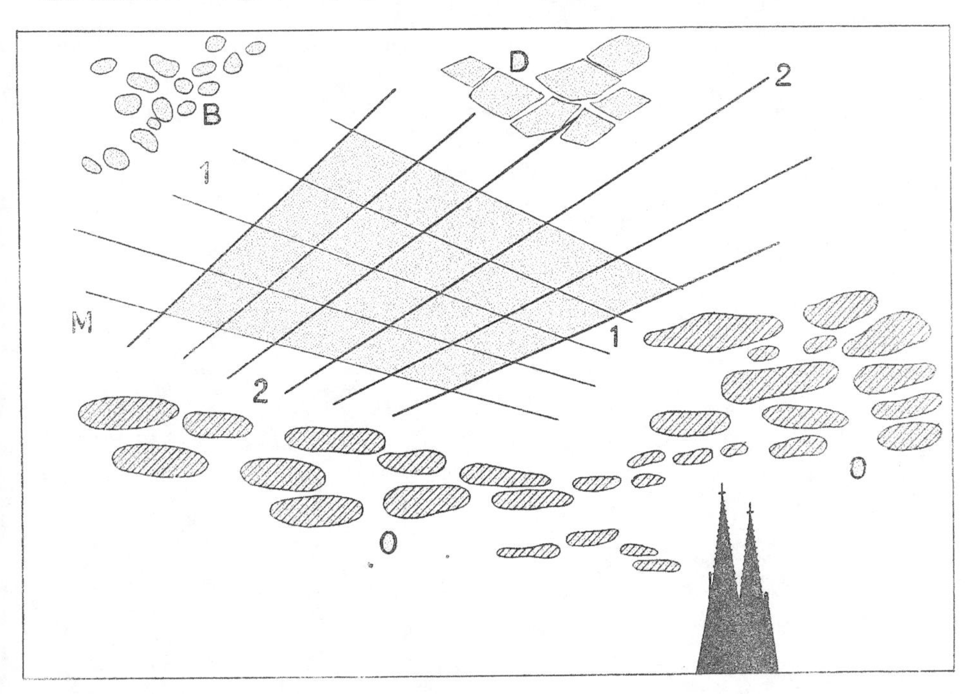

Sheet of Altocumulus at one level (Altocumulus translucidus). — *Code number* **M 3.** —
A regular layer with a structure in two directions (11 and 22). The cloud elements are rather soft, especially at **(M)**; they are in general flattened rounded masses with forms intermediate between a sphere **(B)** and a slab **(D)**. Between them are interstices where the blue sky appears. Although some parts **(OO)** of the layer are rather heavily shaded, its thickness is medium and fairly uniform.

1. Loisel, Juvisy, Sept. 20th 1898.

Delicate Cirrus, not increasing in amount, extensive sheet but not forming a conti-ıous layer (Cirrus filosus). — *Code number* **H 2.** — Cirrus composed of irregularly arranged filaments, entated in various directions ; they have no turned up ends, they are not arranged in sheets or bands, and have no ıdency to fuse together into cirrostratus. They are fairly abundant, but do not increase in amount in any parti-lar direction.

Professor A. Mac Adie. Blue Hill Observatory, Readville, Mass. Sept. 3rd 1924, 11 h 51.

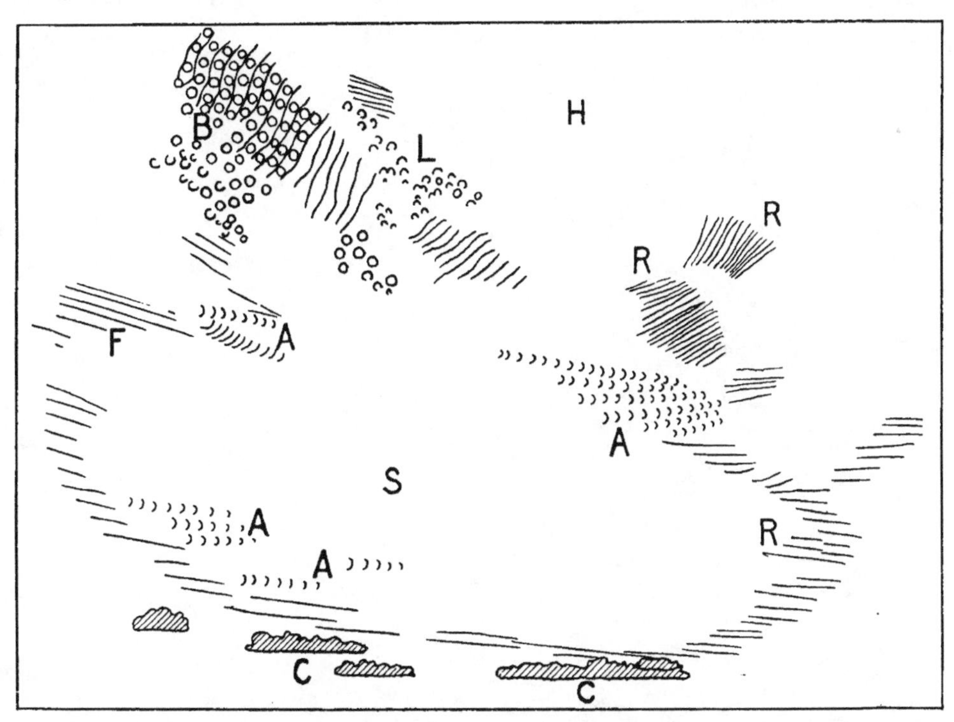

Cirrocumulus predominating, associated with a Cirrus mass. — *Code number* **H 9.** — The cloud bank has a fairly uniform central part **(S)**, perhaps to be classed as cirrostratus. It has very varied characters; at **(B)** small globular masses; at **(L)** a tendency for the formation of rifts; at **(RR)** delicate ripples; at **(AA)** delicate herring-bone structure; at **(H)** a cirriform part; the fibrous nature of the cloud is clearly seen at the edges **(F)**. We are here dealing with a cirrus mass which has changed considerably in the direction of cirrocumulus. At **(CC)** on the horizon are some very flat cumulus clouds.

Code Number	TITLES
L 1	**Cumulus of fine weather** (Cumulus humilis).
L 2	**Heavy Cumulus without anvil** (Cumulus congestus).
L 2	**Cumulus pileus.**
L 2	**Cumulus agitated, rather heavy without anvil.**
L 3	**Cumulonimbus** (Cumulonimbus calvus).
L 3	**Cumulonimbus** (Cumulonimbus incus).
L 3	**Cumulonimbus.**
L 4	**Stratocumulus formed by extension from Cumulus** (Stratocumulus vesperalis. — Melting away of the tops and spreading out of the bases).
L 4	**Stratocumulus formed by the spreading out of Cumulus** (Stratocumulus cumulogenitus. — Melting away of the base, and spreading out of the top).
L 5	**Layer of Stratocumulus** (Stratocumulus translucidus).
L 5	**Layer of Stratocumulus** (Stratocumulus opacus).
L 5	**Layer of Stratus.**
L 6	**Low, ragged, dark grey clouds of bad weather.**
L 7	**Cumulus of fine weather and Stratocumulus.**
L 8	**Heavy Cumulus or Cumulonimbus and Stratocumulus.**
L 9	**Heavy Cumulus or Cumulonimbus with underlying low, ragged clouds of bad weather.**
L 9	**Cumulonimbus with an underlying layer of low, ragged clouds of bad weather, which have a definite roll formation** (Cumulonimbus arcus).
M 1	**Typical thin Altostratus** (Altostratus translucidus).
M 2	**Typical thick Altostratus** (Altostratus opacus).
M 3	**Sheet of Altocumulus at one level** (Altocumulus translucidus).
M 4	**Altocumulus in small isolated lenticular patches** (Altocumulus lenticularis).
M 4	**Altocumulus in small isolated patches more or less lenticular in shape.**
M 4	**Altocumulus in small isolated patches, more or less lenticular in shape and with trailing precipitation** (Altocumulus virga).
M 5	**Altocumulus arranged in parallel bands spreading over the sky.**
M 5	**Altocumulus in a definitely organized layer spreading over the sky** (Altocumulus translucidus).

Code Number	TITLES
M 6	**Altocumulus formed by the spreading out of the tops of Cumulus** (Altocumulus cumulogenitus).
M 7	**Altocumulus associated with Altostratus.**
M 7	**Altocumulus associated with Altostratus.**
M 7	**Altocumulus predominating with parts resembling Altostratus** (Altocumulus opacus).
M 8	**Altocumulus castellatus.**
M 8	**Altocumulus floccus, scattered cumulus-shaped tufts** (Altocumulus floccus).
M 9	**Altocumulus in several patches at different levels, associated with thick fibrous sheets. Chaotic grouping of clouds.**
M 9	**Altocumulus in several patches or layers at different levels, associated with thick fibrous sheets. Chaotic aspect.**
H 2	**Delicate Cirrus, not increasing in amount, extensive sheet but not forming a continuous layer** (Cirrus filosus).
H 3	**Dense Cirrus derived from an anvil** (Cirrus nothus).
H 3	**Dense Cirrus probably derived from anvils** (Cirrus densus).
H 4	**Delicate Cirrus, increasing in amount, filaments in the form of hooks ending in a little claw or upturned end** (Cirrus uncinus).
H 5	**Cirrus and Cirrostratus increasing in amount, but not higher than 45° above the horizon.**
H 6	**Cirrostratus increasing in amount, and more than 45° above the horizon.**
H 7	**Cirrostratus covering all the sky** (Cirrostratus filosus).
H 8	**Cirrostratus not increasing in amount, and not covering all the sky** (Cirrostratus nebulosus).
H 9	**Cirrocumulus predominating, associated with a Cirrus mass.**

Office National Meteorologique, Paris, Aug. 11th 1925, 14 h. 07 looking W, altitude 27°.

Cumulus turbulent, rather heavy, without anvil. — Code number **L 2.** — The rising heads are heaped up, especially at **(C)**, but the clouds are much more broken up than those of plate L 2a; the bases are badly defined **(BB)** and are not definitely horizontal; the cloud masses are more or less broken up **(D)** and there is no vertical symmetry, the cloud tops being drawn out by the wind **(V)**. At **(AA)** there are scattered altocumulus clouds which come from a layer of altocumulus cumulogenitus. At **(S)** is a patch of thick lenticular cirrostratus which comes from the anvil of a cumulonimbus. The presence of these clouds shows the frequent association of this type of cumulus with the debris of middle and high cloud. Such a complex sky with lower, middle and high clouds is coded $C_L = 2$, $C_M = 7$, $C_H = 3$.

F. W. Baker, Yateley, Hants, June 10th 1920, 18 h. 55, looking E.

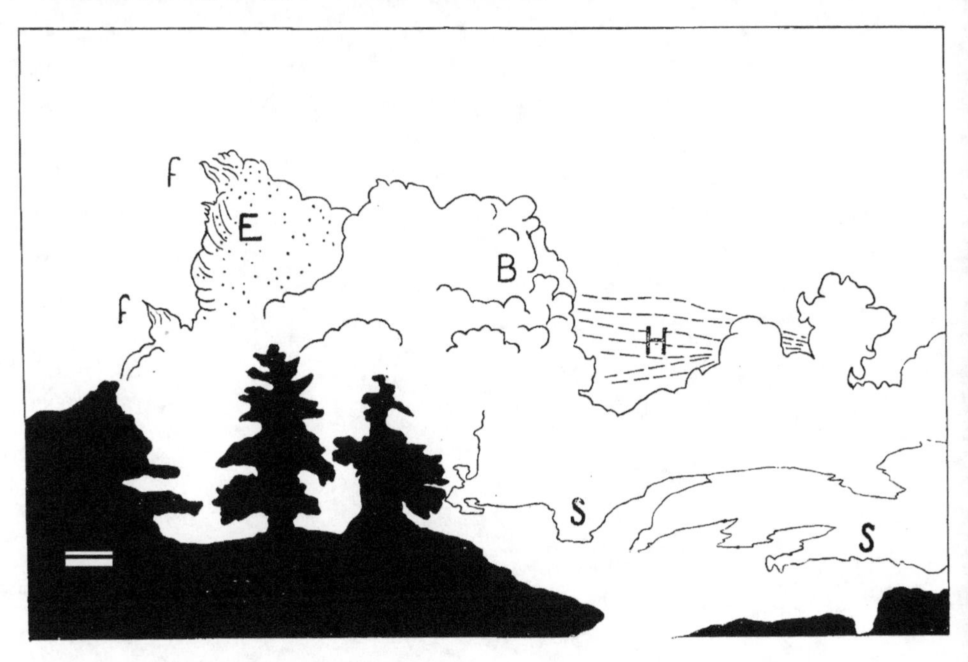

Cumulonimbus (Cumulonimbus calvus).— *Code number* **L 3.** — The tops are just beginning to take on a cirriform structure. At **(FF)** the cloud is " smoking, " a sign that it is in active growth. An anvil is beginning to form at **(E)**. The clear cut cumulus structure at **(B)** begins to soften. and the hard " cauliflower " domes dissolve and give place to a fibrous structure; this state is generally rather ephemeral, being one of transition, and the complete cirrus anvil is soon formed. At **(SS)** are dark fractocululus clouds of the base. At **(H)** are independent cirrus clouds.

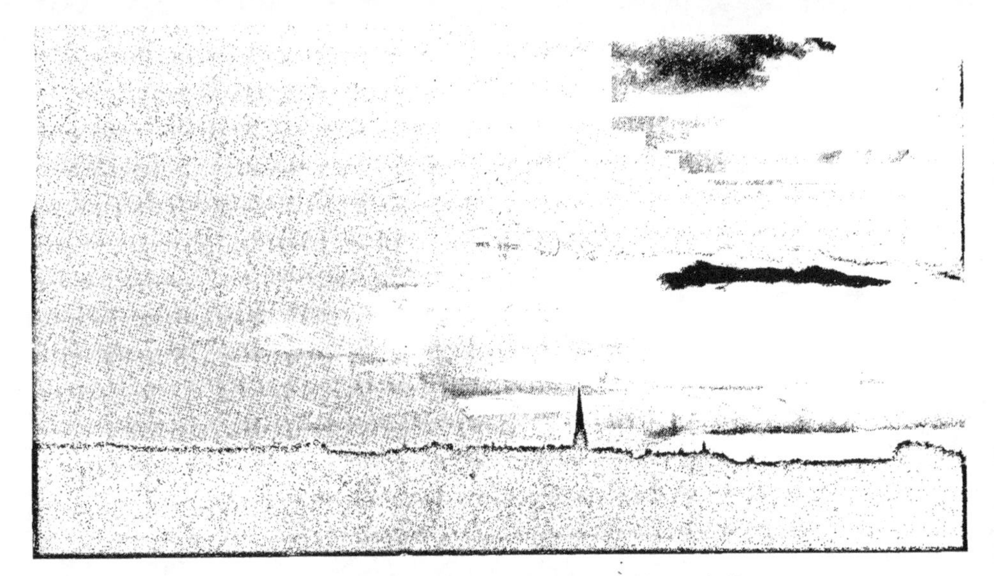

Office National Météorologique, Paris, Sept. 24th 1925, 15 h. 08, looking N, altitude 13°.

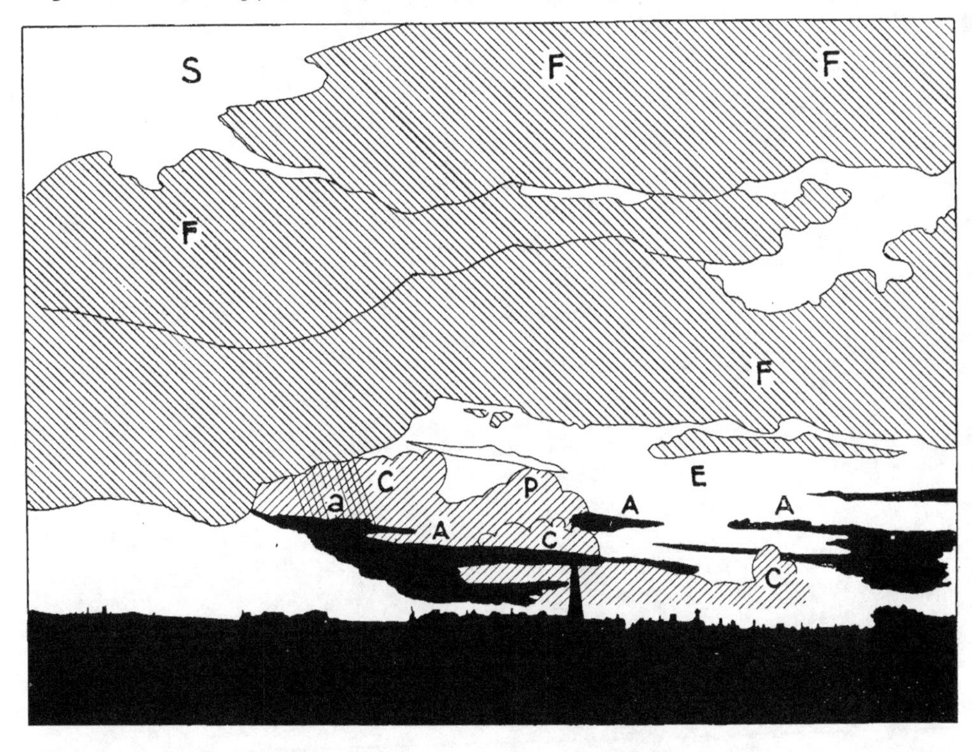

Heavy Cumulus or Cumulonimbus with underlying low, ragged clouds of bad weather. — *Code number* **L 9.** — The rounded cumulus at **(CC)** is seen rising above distant banks of stratocumulus or altocumulus **(AA)**. At **(P)** a veil cloud is lifted up by the top of a cumulus. At **(S)** is the base of a cumulonimbus (resembling nimbostratus) with low ragged clouds below **(FF)**, fractostratus rather than fractocumulus ; these seem to encroach on the base of the cumulonimbus. There is a shower at **(a)** and a rift at **(E)**.

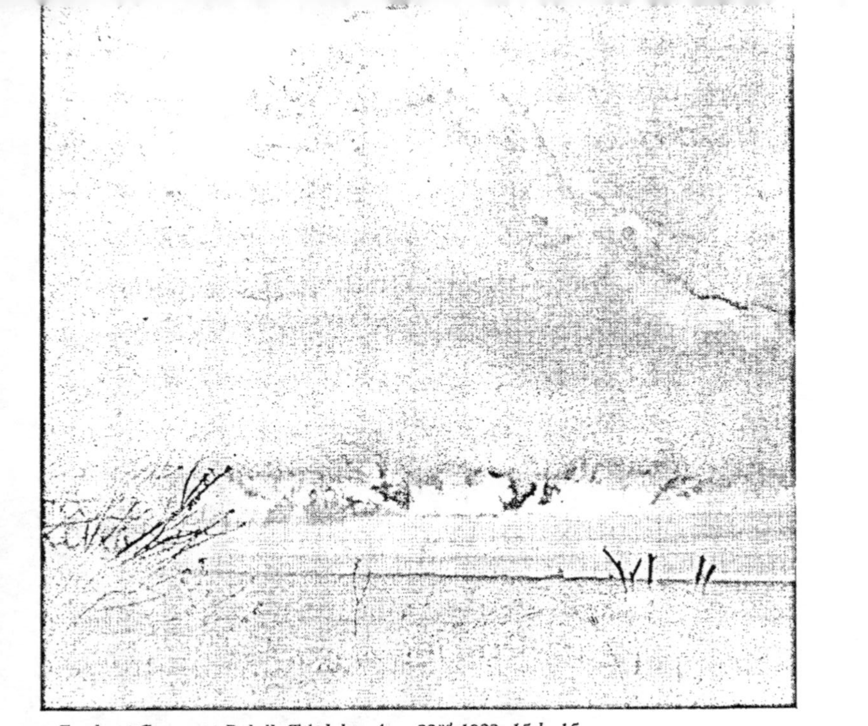

Fundació Concepció Rabell, Tibidaba, Apr. 22nd 1923, 15 h. 15.

Cumulonimbus with an underlying layer of low, ragged clouds of bad weather, which have a definite roll formation (Cumulonimbus arcus). — *Code number* **L 9.** — The tops of the cumulonimbus are hidden for the base covers the zenith (whereas in photograph L 3c pl. 6, the anvil is still visible although the cloud approaches the zenith). The base consists of a grey veil **(NN)**, closely resembling nimbostratus. Lying under it are small, low, ragged clouds **(FF)**, and a rather marked cloud mass of roll form (arcus). Notice rounded cumulus heads at **(BB)**, a turbulent zone **(T)**, and at the base ragged masses **(EE)**, more or less turbulent, which stand out dark against a relatively lighter background of cloud.

F. W. Baker, Blackwater, Hants, Sept. 30th 1923, 7 h. 16, looking W.

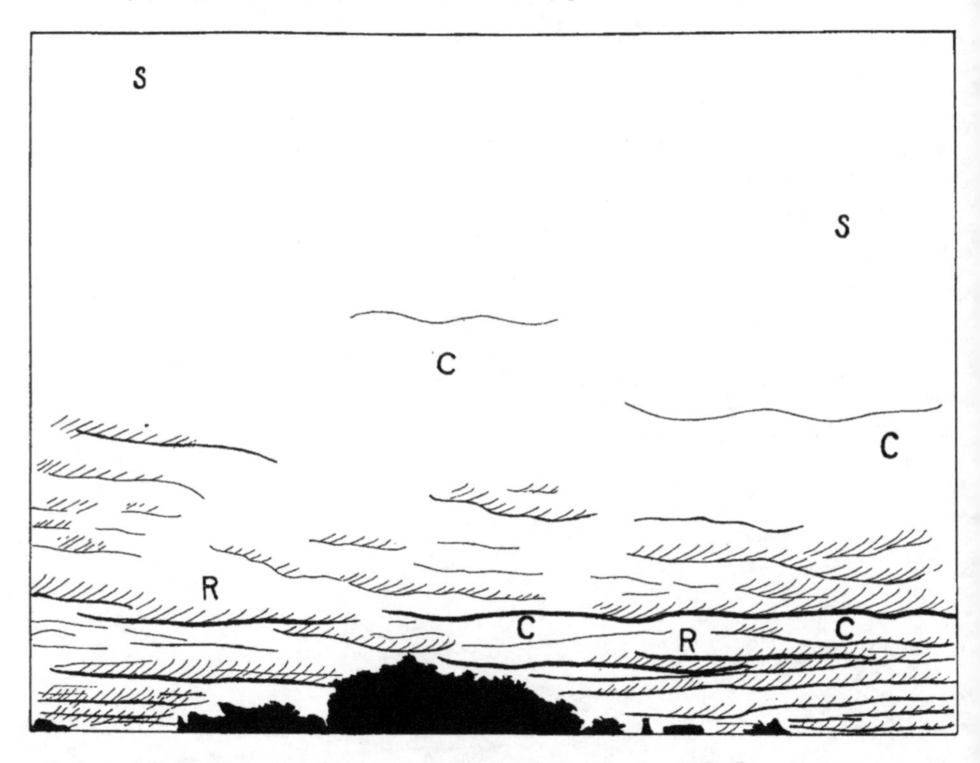

Layer of Stratocumulus (Stratocumulus opacus). — *Code number* **L 5.** — A rather indistinct layer showing some light and shade contrasts near the zenith **(CS)** and a roll structure **(RR)** near the horizon where it is accentuated by perspective, showing that the layer has a more or less regular structure. Transition into Stratus.

Office National Météorologique, Paris, Nov. 30th 1925. 13 h. 50, looking ESE. altitude 15°.

Low, ragged, dark grey clouds of bad weather. — Code number **L 6.** — These low clouds **(CC)** show up very dark against the relatively lighter background of altostratus or nimbostratus which appears in places **(S)**, especially near the zenith. On the horizon the clouds close up owing to perspective, and appear as large irregular rollers **(RR)**. The clouds, which show some relief with rounded outlines **(aa)**, are rather fractocumulus than fractostratus.

F. W. Backer, Blackwater, Hants, Sept. 29th 1923, 13 h. 04, looking N.

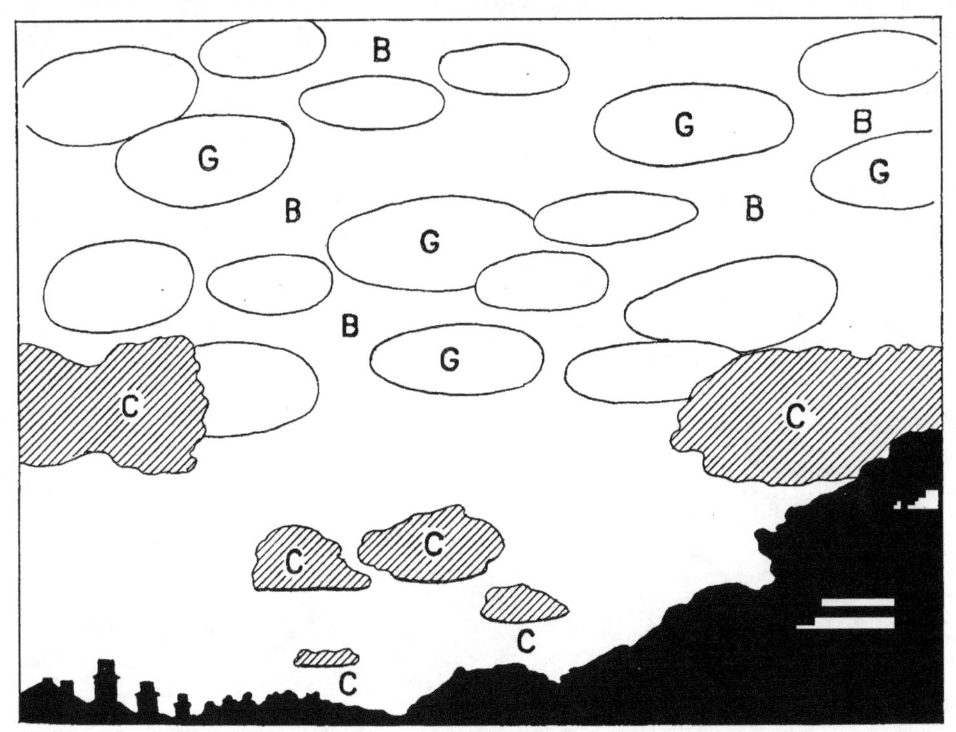

Cumulus of fine weather and Stratocumulus. — *Code number* **L 7.** — Stratocumulus in large soft masses which are almost without shadow, and hence are thin. The elements are flattened rounded masses **(GG)**, and between them blue sky **(BB)** is seen. By itself the clouds would be coded $C_L = 5$ (cf. L 5a. pl 9), but feebly developed cumulus clouds **(CC)**, which by themselves would be coded $C_L = 1$, lie below the cloud sheet, their tops nowhere reaching it ; this shows that the layer is independent and is not formed from a spreading out of the tops of the cumulus.

C. J. P. Cave, Bosham, Sussex, September, 1924

Heavy Cumulus, or Cumulonimbus, and Stratocumulus. — *Code number* **L 8.** — At **(RR)** is a corrugated sheet of cumulonimbus which if it were alone would be coded $C_L = 5$ (cf. L 5b. pl 10). At **(CC)** there is lower cumulus which penetrates the stratocumlus at **(PP)**. It is a definite case of penetration without any transition of one cloud form into the other. The stratocumulus is independent of the cumulus and does not originate in a spreading out of the latter ; but the convection is sufficiently powerful for the cumulus to be strongly developed, so that it reaches and penetrates the pre-existing layer ; perhaps the convection clouds are cumulonimbus, but since their tops are invisible this cannot be determined.

Fundació Concepció Rabell (M. Pulvé), Barcel.na, Nov. 9th 1923, 8 h. 10.

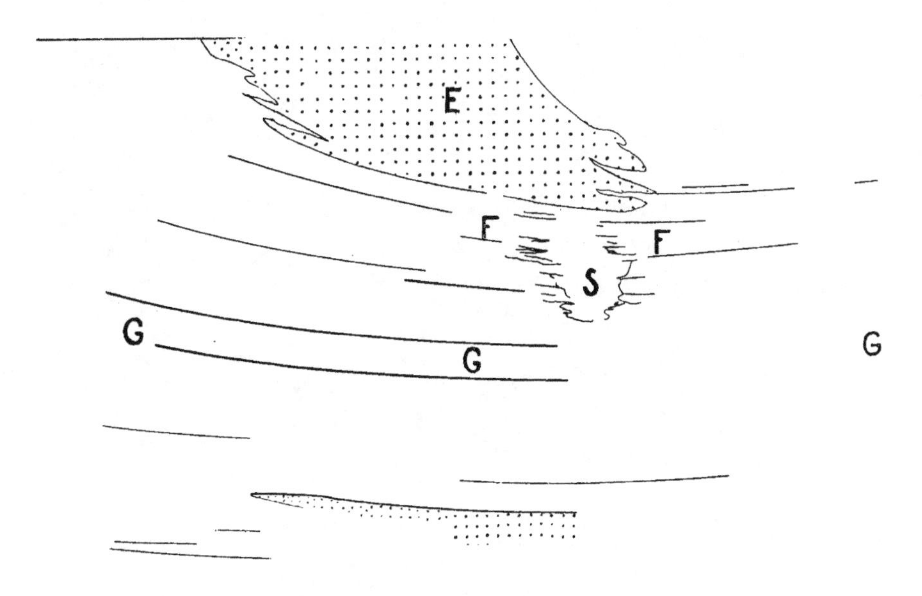

Typical thick Altostratus (Altostratus opacus).— *Code number* **M 2.** — The layer covers whole sky, but its thickness varies very much ; at **(GG)** it is very dark and thick; at **(E)** thin. The sun still s as a light patch **(S)** round which the fibrous structure of the cloud **(FF)** is seen. But it is clear that if the sun behind the part **(GG)** it would be completely hidden.

C. J. P. Cave, Petersfield, Hants, April 21st 1924, about 8 h., looking E.

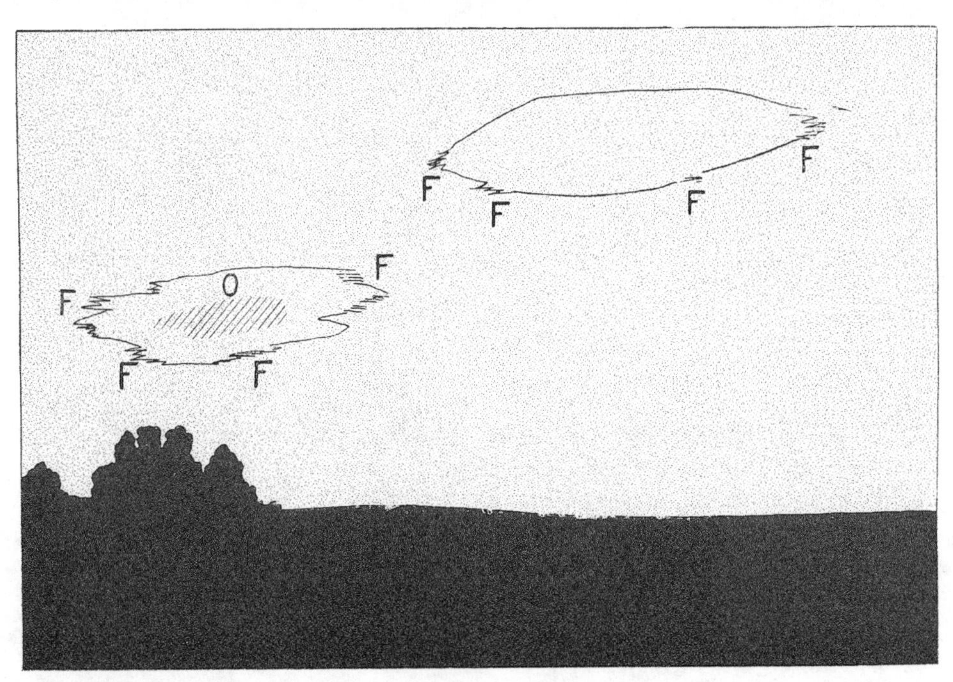

Altocumulus in small isolated lenticular patches (Altocumulus lenticularis). — *Code number* **M 4.** — This is a case of typical lenticular altocumulus, with well defined lens shaped masses of a dazzling white, and capable of producing beautiful irisations. Although the edges of the patches show a shredded and fibrous outline, and though the patch on the right has no shadow, the clouds should be classed as altocumulus and not cirrocumulus ; these patches being quite alone in the sky, there is no connection between them and any cirrus mass ; also the patch on the left, obviously of the same kind as the other, has strong shading **(O)**.

Office National Météorologique, Paris, Nov. 17th 1926, 14 h. 55, looking WSW, altitude 35°.

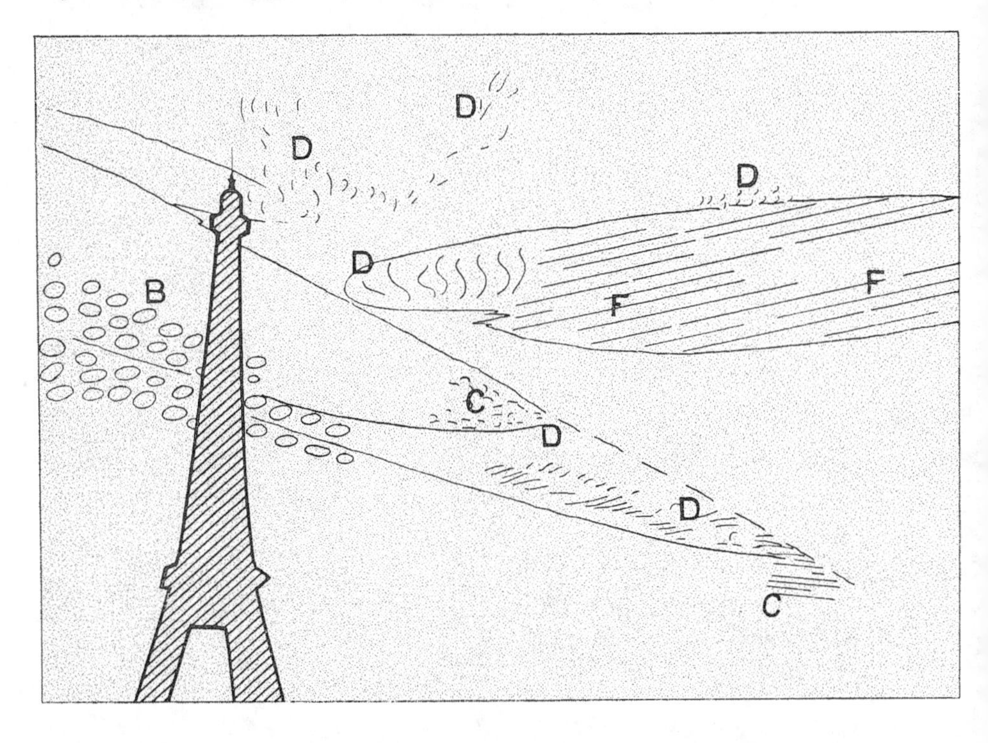

Altocumulus in small isolated patches more or less lenticular in shape. — *Code number* **M 4.** — The lenticular tendency in the patches is obvious. The details show a complex structure, from a typical globular shape **(B)** to the small tessellations and ripples of cirrocumulus **(C)**, and even to a fibrous appearance **(FF)**. There are probably at least two layers of cloud: the upper one composed of very delicate and light patches **(C, FF)**, the lower typical altocumulus (at **B** and below) strongly shaded. The cloud patches are changing rapidly ; note that the cloud at **(DD)** is in process of dissolution.

eteorologisch-Magnetisches Observatorium, Potsdam, May 28th 1900, 17 h., looking WNW, altitude 25°.

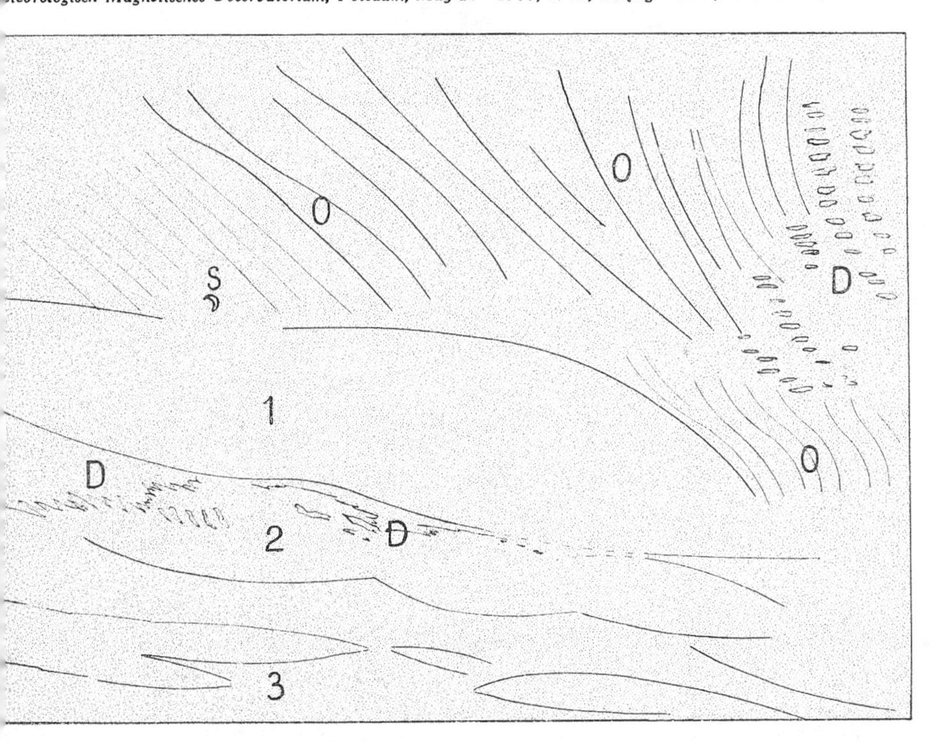

Altocumulus arranged in parellel bands spreading over the sky. — *Code number* **M 5.** —
e great parallel bands (1, 2, 3) are made up of roughly lenticular elements fused together longitudinally, and are ly thick (strong shading). At **(OO)** may be seen a very pronounced wave structure at right angles to the general ection of the bands. In certain places **(DD)** on the edges of the bands there is a partial evaporation, but the total ount of cloud is large. At **(S)** is the sun (partially eclipsed by the moon).

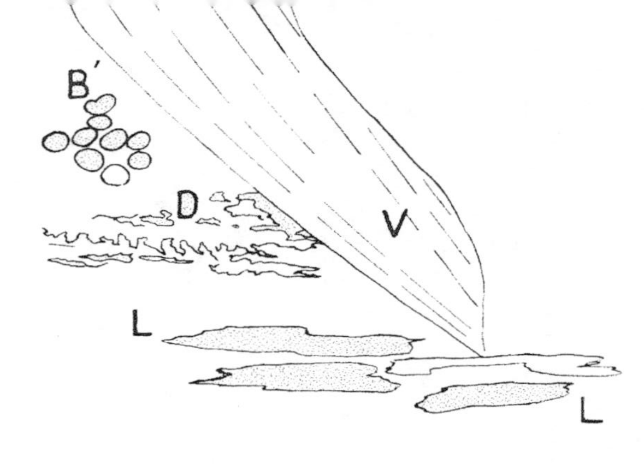

Meteorologisch-Magnetisches Observatorium, Postdam, May 25th 1904. 12 h., looking WSW, altitude 18°.

Altocumulus in small isolated patches, more or less lenticular in shape and with trailing precipitation (Altocumulus virga). — *Code number* **M 4.** — The cloud elements of the principal patch are very irregularly arranged; it seems of considerable extent because it is at a high elevation above the horizon, but the total amount of cloud in the sky is small. At **(BB')** can be seen the typical globular structure of altocumulus; the elements **(B)** seem at first sight to be fairly large, but in reality each is made up of smaller globular masses fused together. Trailing wisps of rain **(VV)** called virga, are falling from the part of the cloud near the zenith, but they evaporate before reaching the ground. These virga are of rather exceptional size. At **(DD)** it is evident that the patch is in process of evaporation. Lower down near the horizon the patches **(L)** show a more or less lenticular shape in perspective.

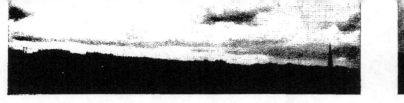

Office National Météorologique, Paris, Sept. 25th 1925, 10 h. 46, looking N, altitude 13°.

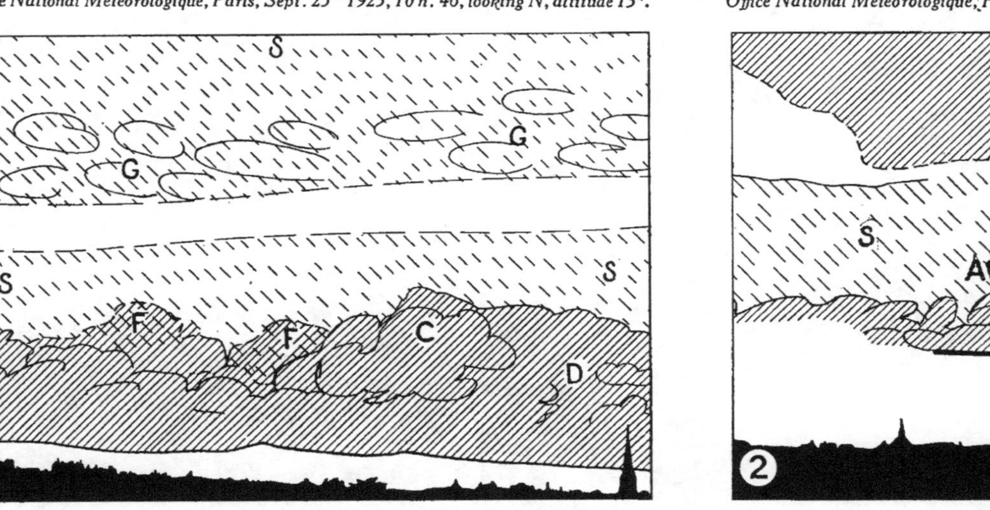

Office National Météorologique, Paris, Sept. 25th 1925, 10 h. 46, looking ENE, altitude 10°.

Altocumulus formed by the spreading out of the tops of Cumulus (Altocumulus cumulogenitus). — Code number **M 6.** — The rather rounded cumulus is becoming flatter **(II)** (photograph 2), the summits spread out and end by forming a layer of altocumulus **(SS)**. The phenomenon is especially well seen at **(A)** (photograph 2). The structure is well shown in photograph 2; large soft rounded masses **(GG)** which are not very thick have no shadows, and have interstices where the blue sky shows through.

C. J. P. Cave, Petersfield, Hants, Sept. 27th 1923, 18 h., looking W.

Altocumulus associated with Altostratus. — *Code number* **M 7.** — The veil of altostratus is very variable in thickness ; at **(VV)** especially, are thicker and darker parts, at **(EE)** thinner and lighter parts. The altocumulus clouds which are below the altostratus consist of elongated and more or less lenticular patches **(LL)**, or of globular masses **(B)**. There are no low clouds. During the subsequent development of these clouds there were only traces of rain.

M. Boissonas, Geneva, Sept. 25th 1923, 18 h., looking SW.

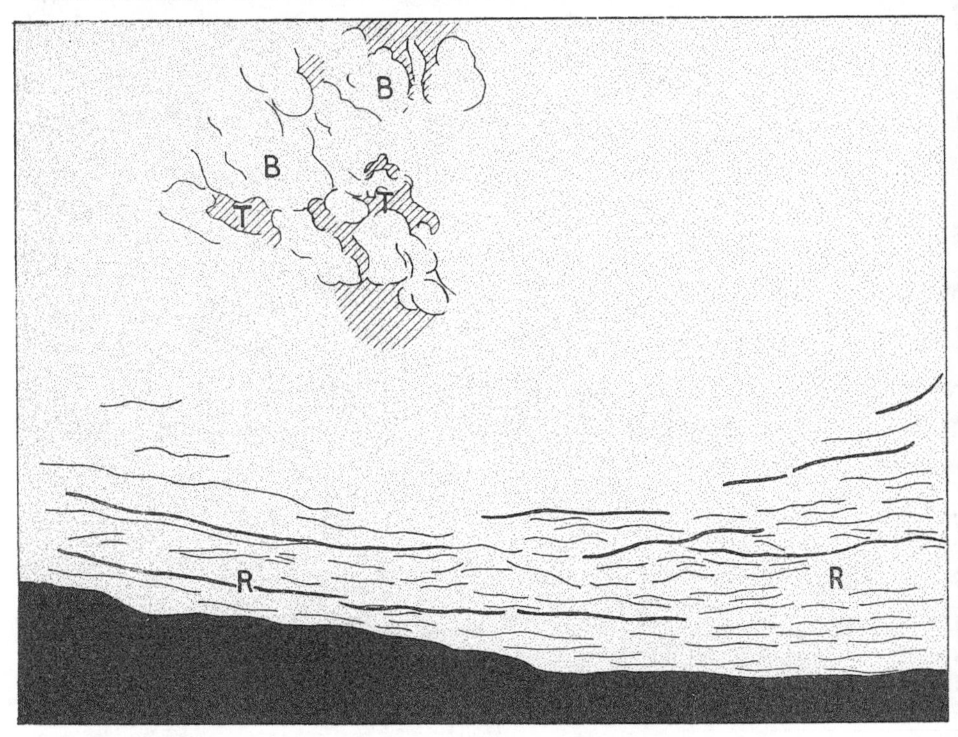

Altocumulus associated with Altostratus. — Code number **M 7.** — Two cloud layers can be seen. The upper layer is composed of globular masses of altocumulus **(BB)** between which a good deal of blue sky is seen **(TT)**. The lower layer is composed of fragments of a corrugated veil **(RR)**. This cloud association is sometimes called altocumulus duplicatus. There is an absence of low cloud. This cloud formation can only cause a very light rain.

F. W. Baker, Blackwater, Hants.

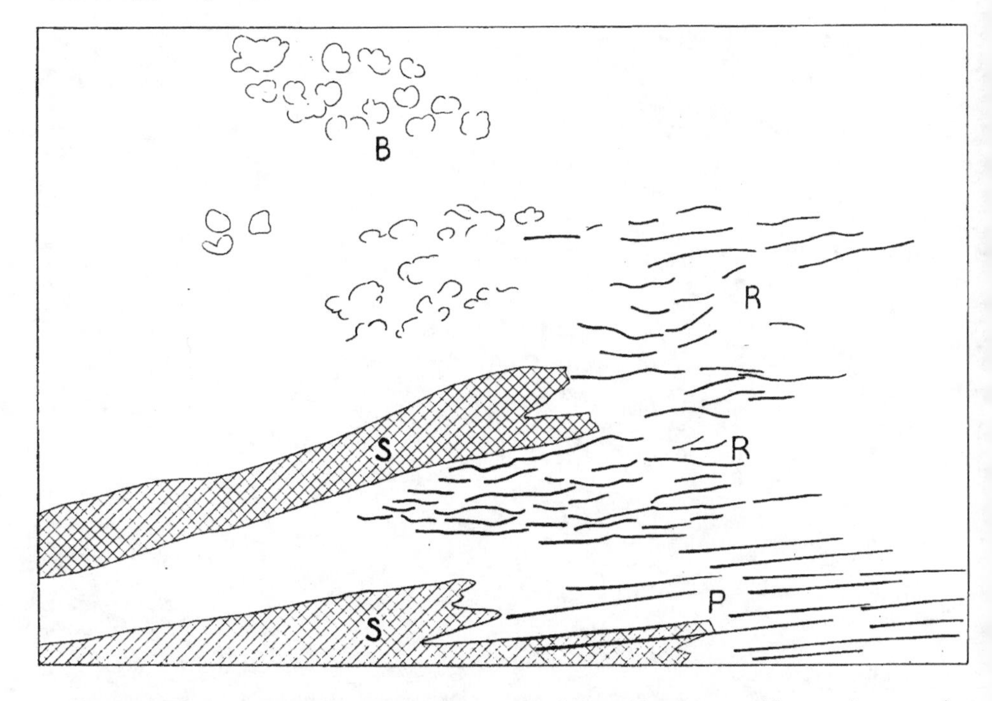

Altocumulus predominating with parts resembling Altostratus (Altocumulus opacus). — Code number **M 7.** — Thick and compact layer of altocumulus with a rippled appearance. At **(B)** the globular structure is apparent ; at **(RR)** ripples ; at **(P)** the ripples are closer owing to the effect of perspective, and make more or less parallel folds. Notice that the cloud elements, both globular masses and ripples, appear in real relief, the effect not being due to differences of transparency. At **(SS)** are nearly uniform dark parts, tending towards altostratus or nimbostratus. Such clouds can only give light rain.

C. J. P. Cave, Bosham, Sussex, Aug. 25th 1926, about 15 h., looking S

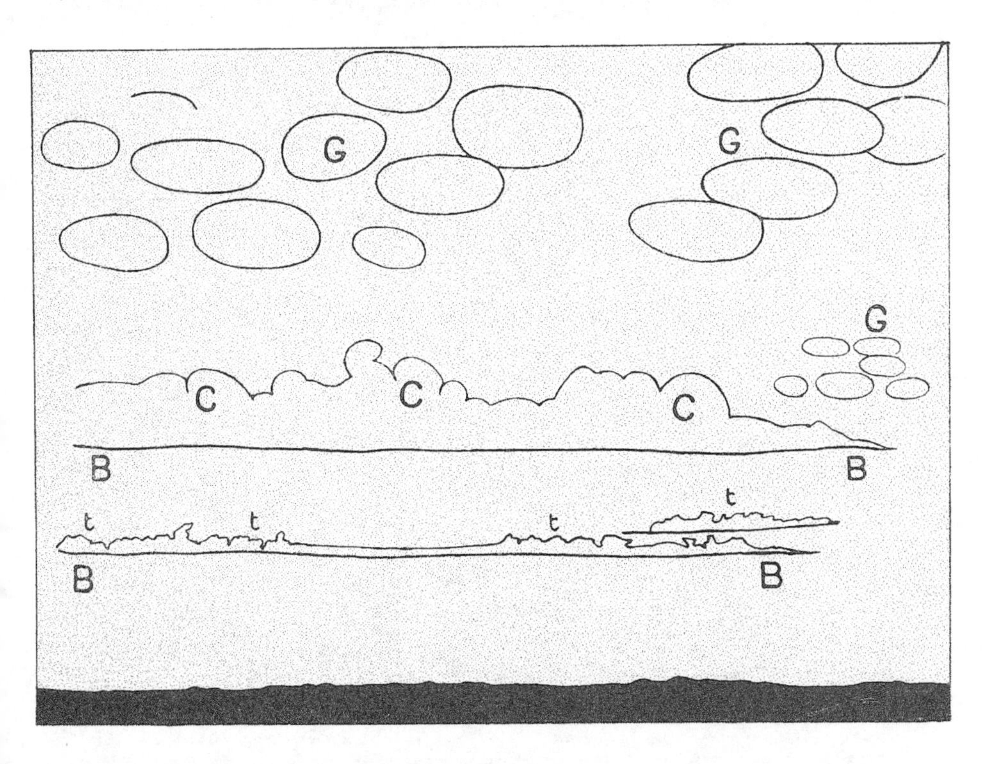

Altocumulus castellatus. — *Code number* **M 8.** — The clouds near the zenith, seen from below, show the typical appearance of altocumulus, with soft rounded masses **(GG)**. Seen near the horizon they are arranged in lines **(BB)** and show a pronounced cumuliform development **(CC)**. At **(tt)** are little turrets arranged in lines, and resting on a common horizontal base **(BB)** which is characteristic of the sub class "castellatus." These clouds preceed thunderstorms.

C. J. P. Cave, Petersfield, Hants, Aug. 5th 1923, 15 h. 30, looking NE

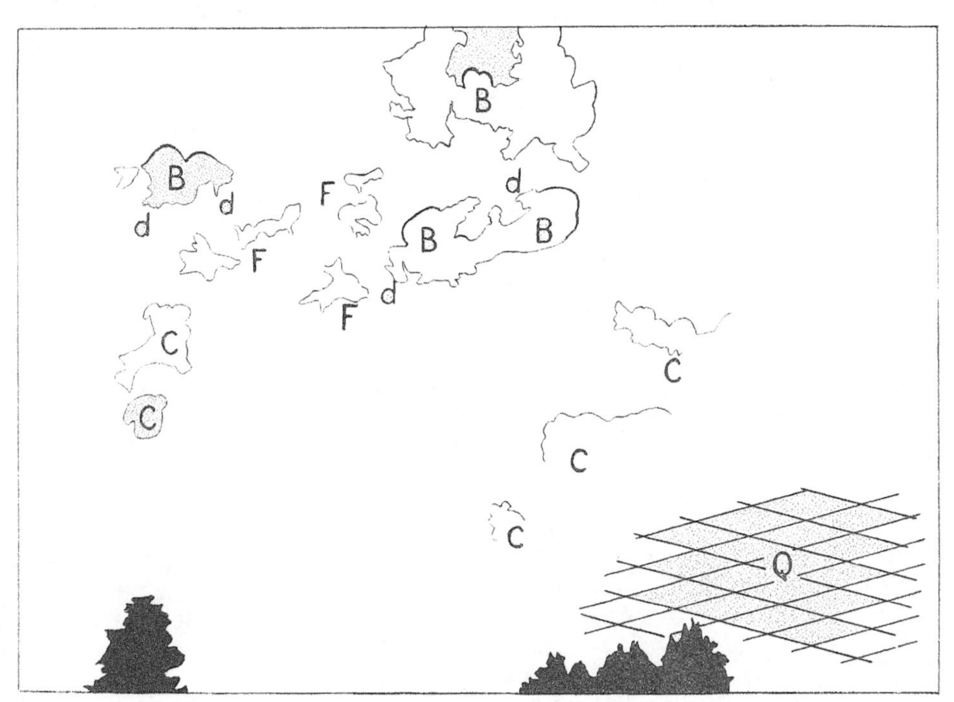

Altocumulus floccus, scattered cumulus-shaped tufts. — *Code number* **M 8.** — The cloud elements **(CC)** resemble small fractocumulus clouds, or cumulus without bases, and have no characteristic shadows. Some elements **(FF)** are quite ragged and broken up ; others are partially so **(dd)** ; others **(BB)** are ill developed globular masses, rather white and, resembling cumulus heads. At **(Q)** the elements are arranged in a series of diagonal lines, the allignment being heightened by perspective. These clouds preceed thunderstorms.

Office National Meteorologique, Paris, July 8th 1922, 9 h. 40, looking SE.

Altocumulus in several patches at different levels, associated with thick fibrous sheets. Chaotic grouping of clouds. — *Code number* **M 9.** — At **(CC)** are rounded cumulus clouds remarkable at so early an hour, (9.40). At **(B)** is a patch of altocumulus having an irregular globular structure and tending towards the " floccus " type at **(f)**. At **(EE)** is a remarkabale group of these altocumulus clouds. At **(FF)** are little tufts with a cirriform appearance, but at **(A)** there is a compactness characteristic of altocumulus. At **(DD)** there is a large layer of high fibrous cloud, which is difficult to classify. The clouds appear chaotic but not ragged. Thundery appearance.

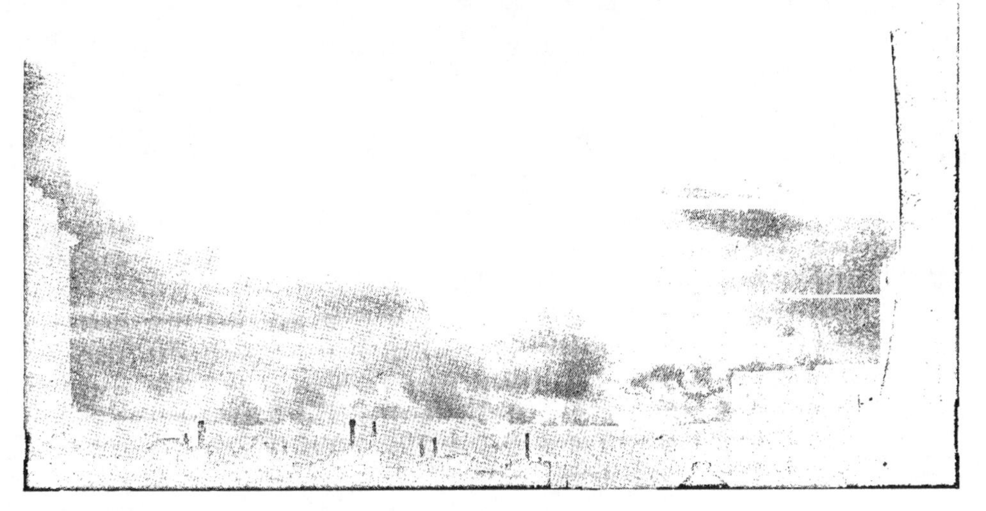

Office National Meteorologique, Paris, Aug. 11th 1925, 8 h. 55, looking W altitude 27°.

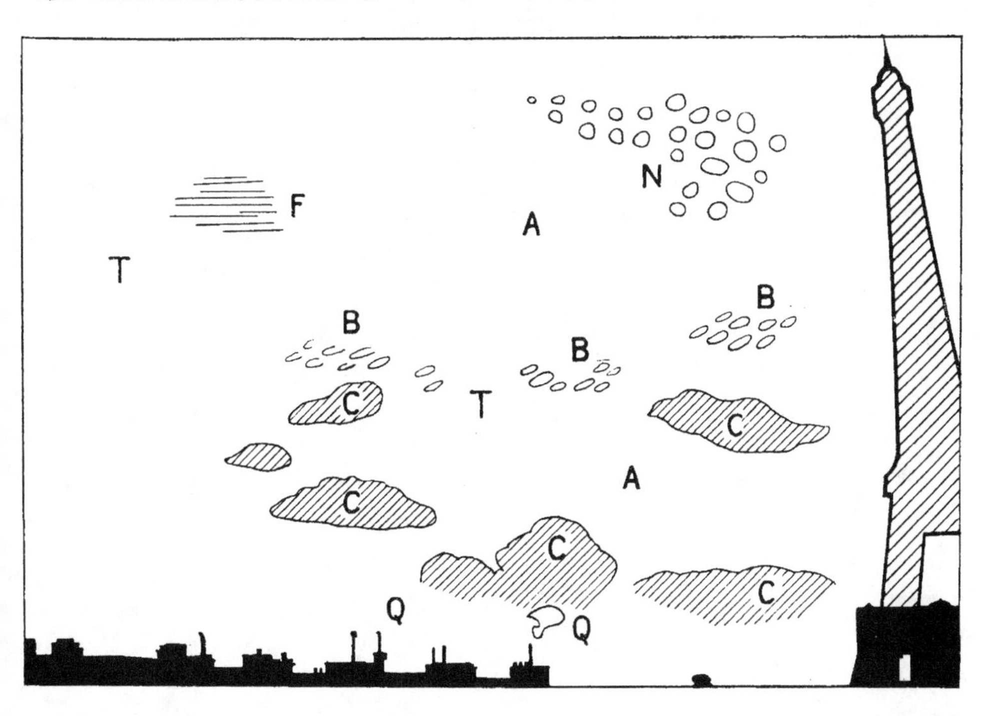

Altoscumulus in several patches or layers at different levels, associated with thick fibrous sheets. Chaotic aspect. — *Code number* **M 9.** — At **(CC)** are cumulus clouds in spite of the early hour (8.55) ; these are markedly rounded in places **(QQ)**. At **(AA)** are clouds at different levels, as is well seen at **(N)**. Here the clouds at the low level are altocumulus, showing dark, with globular structure **(BB)** ; at the higher level is a lighter. but thick fibrous sheet of doubtful species ; its fibrous structure appears in places **(F)** ; there are rifts where blue sky shows **(TT)**. The clouds are chaotic but not ragged ; they are "heavy" and "stagnant" (absence of wind). Thundery appearance.

Meteorologisch-Magnetisches Observatorium, Potsdam, Sept. 29th 1911, 15 h. 35, looking NW.

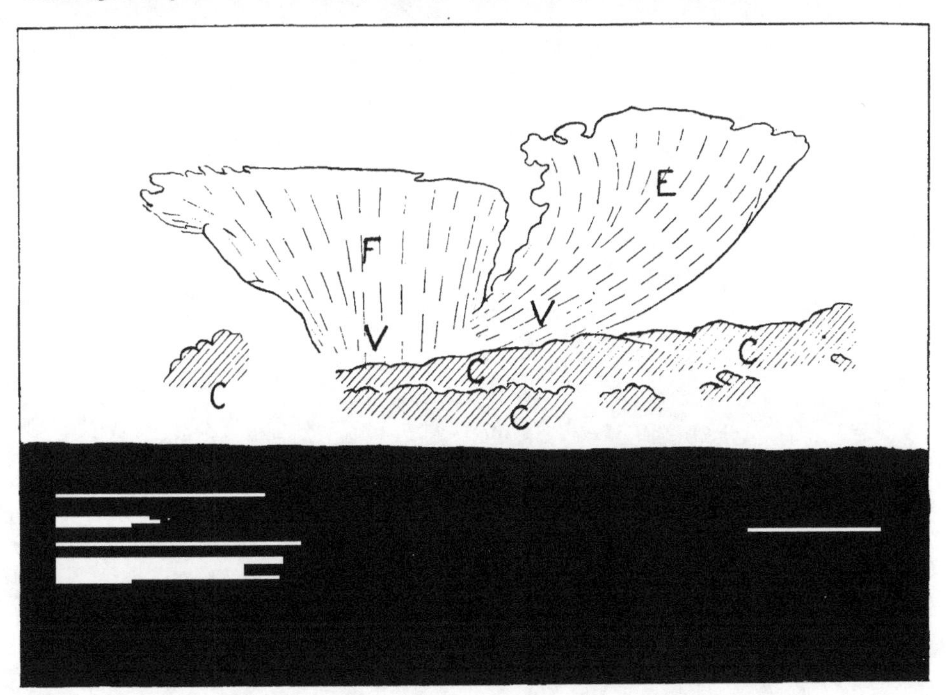

Dense Cirrus derived from an anvil (Cirrus nothus). — Code number **H 3.** — At **(E)** and at **(F)** are two cirrus masses still having the shape of an anvil, particularly at **(F)**. They are dense, having shadows. Below these masses **(E)** and **(F)** are soft streamers **(VV)** " virga, " showers of snow which do not reach the ground. At **(CC)** slightly rounded cumulus masses are still visible, but on the whole they are degenerate.

Ebro Observatory, Tortosa, June 20th 1911. 9 h. 40, looking W.

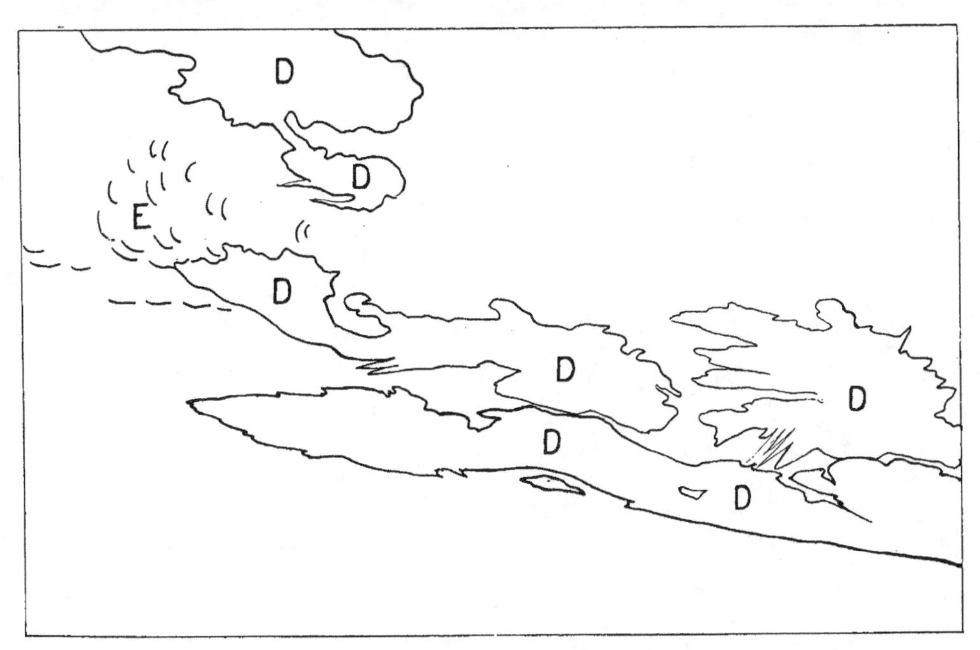

Dense Cirrus probably derived from anvils (Cirrus densus). — *Code number* **H 3.** — The cirrus forms little masses whose density is especially noticeable at **(DD)**. At **(E)** is an appearance of spindrift characteristic of a thundery tendency. These clouds are formed in all probability from the growth of cumulonimbus clouds which have lost their rounded parts: they have no doubt been formed some considerable time for the anvil shape is completely lost.

C. J. P. Cave. Petersfield, Hants, Oct. 31st 1923, 15 h. 30, looking S.

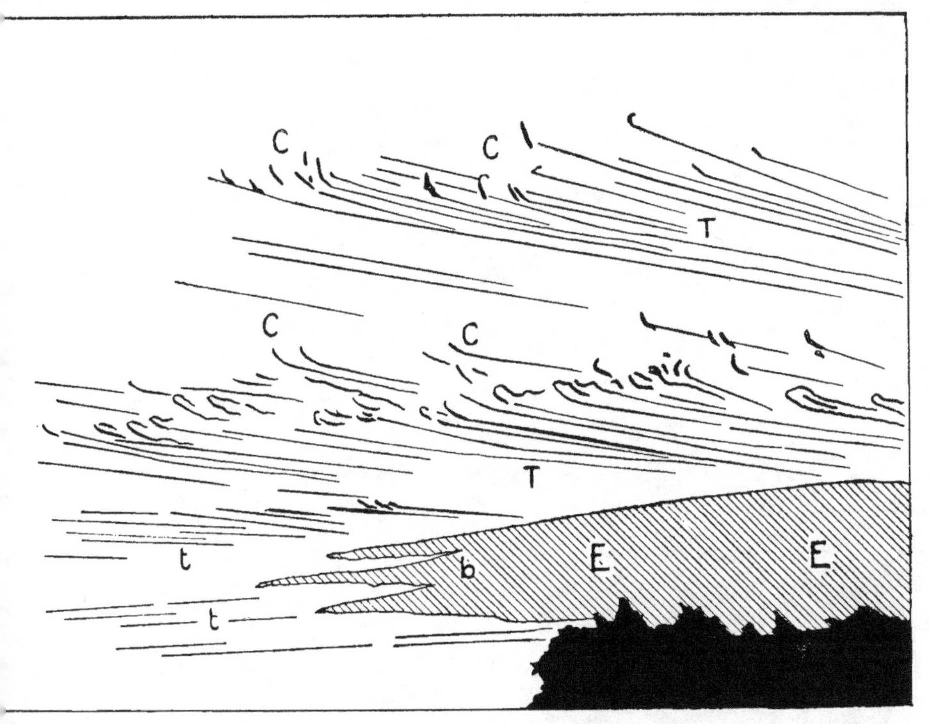

Delicate Cirrus, increasing in amount, filaments in the form of hooks ending a little claw or upturned end (Cirrus uncinus). — Code number **H 4.** — The upturned ds are well seen at **(CC)** and the filaments at **(TT)**. In the lower part of the photograph the cirrus has not the uncinus aracter, but filaments **(tt)** parallel to those at **(TT)** can be distinguished. The cloud bank **(EE)** is strongly shaded ; **(b)** the cloudlets have a globular form, and the whole cloud belongs to a lower level (altocumulus). There is a finite tendency for an increase of cloudiness as the horizon is neared, but the clouds do not even here form a con-uous layer. The clouds are advancing towards the observer, and the sky will become more overcast and thus ere will be in time an increase of cloudiness.

Meteorologisch-Magnetisches Observatorium. Potsdam, June 20th 1900, 7 h. 39, looking NW, altitude 2°.

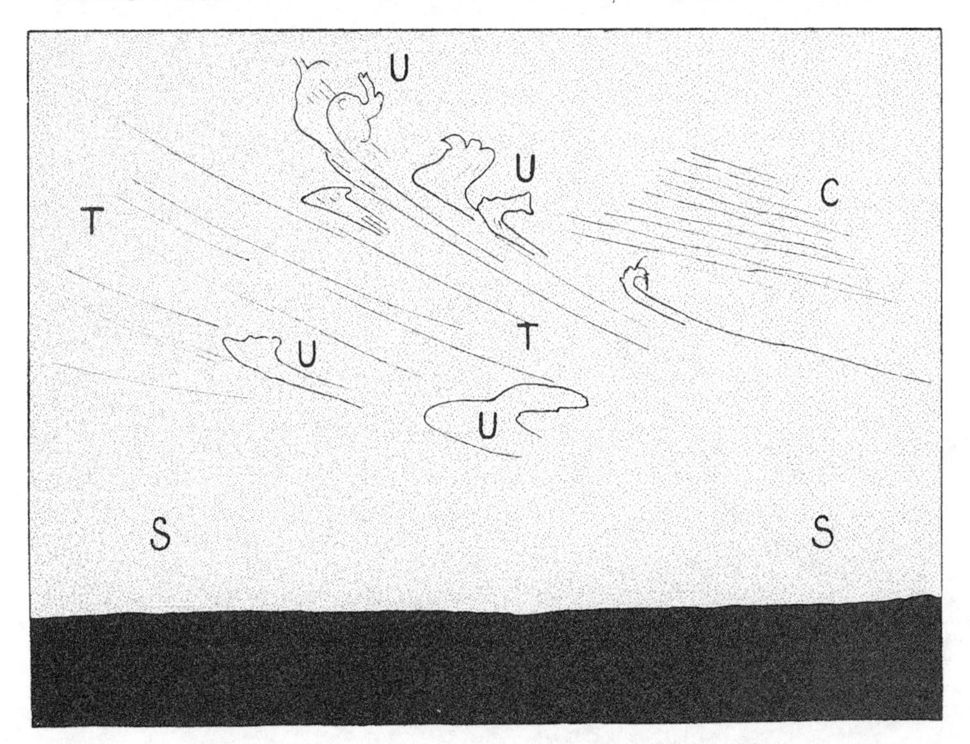

Cirrus and Cirrostratus increasing in amount, but not higher than 45° above the horizon. — *Code number* **H 5.** — The top of the cloud structure is made up of straight lines of cirrus (**TT**), some of them ending in little tufts or points (**UU**). Nearer the horizon the cirrus strands are fused together into a nearly uniform veil of cirro stratus (**SS**). At **C** are some negligeable cirrocumulus ripples. There is a characteristic increase of cloudiness in the sheet as the horizon is neared, and in time also there will probably be a general increase in cloudiness. The clouds are rising from the horizon towards the observer. The front of the sheet of cirrus and cirrostratus is not at a high altitude; it is certainly lower than 45°.

Office National Meteorologique, Paris, Feb. 15th 1926, 14 h. 05, looking S, altitude 25°.

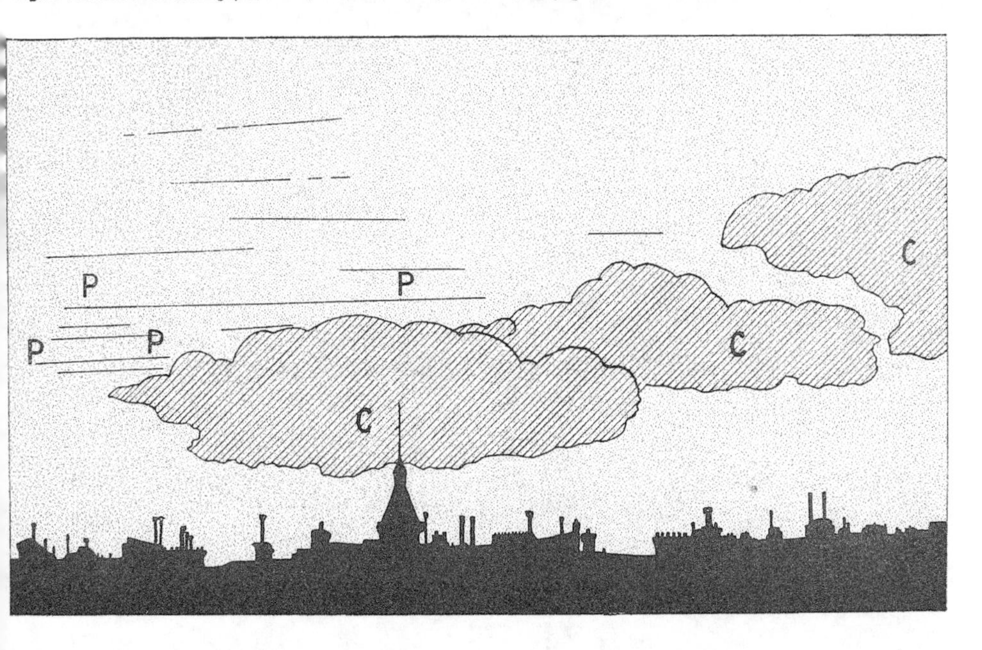

Cirrostratus covering all the sky (Cirrostratus filosus). — *Code number* **H 7.** — The whole sky is covered with a thin and almost uniform sheet of cirrostratus; a rather confused structure of parallel bands can only just be made out **(PP)**. At **(CC)** are cumulus clouds whose shading is accentuated by the upper cloud sheet; the cumulus is very much flattened, as is often the case below a high or middle cloud layer.

G. A. Clarke, Aberdeen, April 1st 1917, 12 h. 30, looking W.

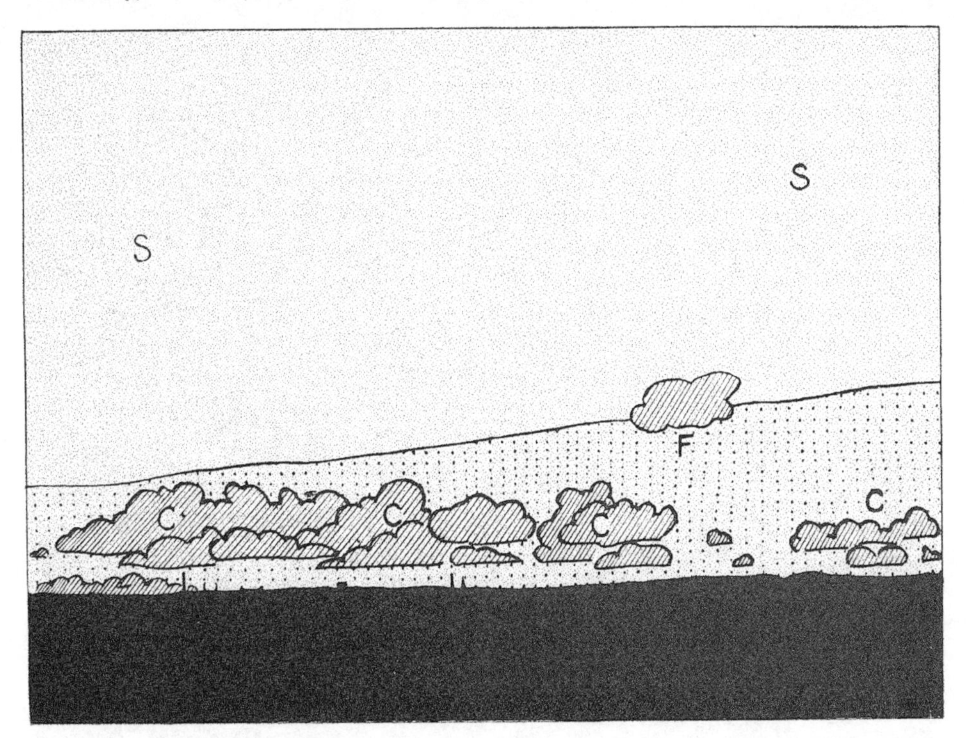

Cirrostatus not increasing in amount, and not covering all the sky (Cirrostratus nebulosus). — *Code number* **H 8.** — At **(SS)** is a sheet of rather dense cirrostratus (the density is exaggerated in the photograph) which covers the whole sky from the horizon behind the observer to the edge which is quite sharp. At **(CC)** are small degenerate cumulus clouds ; a Fractocumulus is to be found in **(F)**. One may suppose that the clear part of the sky does not increase in extent, that is that the sheet **(SS)** moves at right angles to the line of sight.